3D Studio VIZ®
Tutorials

3D Studio VIZ® Tutorials

Daniel Douglas

autodesk®
press

Australia • Canada • Mexico • Singapore • Spain • United Kingdom • United States

THOMSON LEARNING™

autodesk® press

3D Studio VIZ® Tutorials
by Daniel Douglas

Business Unit Director:
Alar Elken

Executive Editor:
Sandy Clark

Acquisitions Editor:
James DeVoe

Development Editor:
John Fisher

Editorial Assistant:
Jasmine Hartman

Executive Marketing Manager:
Maura Theriault

Marketing Coordinator:
Paula Collins

Executive Production Manager:
Mary Ellen Black

Production Manager:
Larry Main

Production Editor:
Stacy Masucci

Art and Design Coordinator:
Mary Beth Vought

Cover Image:
Brucie Rosch

Library of Congress Cataloging-in-Publication Data
Douglas, Daniel.
 3D Studio VIZ Tutorials/Daniel Douglas.
  p. cm.
 ISBN 0-7668-2869-7
 1. Computer graphics. 2. 3D Studio
 I. Title.
 T385 .D685 2000
 006.6'933-dc21
 00-045344

CONTENTS

INTRODUCTION

Welcome to 3D Studio VIZ Tutorials.

As you would expect from the title, this is not a reference book that you will read passively. This is a book of tutorials, written for anyone needing to depict natural and built environments with 3D Studio VIZ: architects, engineers, landscape architects, interior designers, product designers, and so on.

No prior knowledge of VIZ or of CAD in general is needed to use this book. There are occasional references to connections between VIZ and AutoCAD, but these references are not part of the workflow, so if you do not use AutoCAD, disregard them. There is not a chapter concerning VIZ and AutoCAD interoperability. Mastery of that topic involves more reading than practice, and this is a book of exercises.

The book is written with the same approach the author uses in the classroom; very hands-on, with an emphasis on reiteration of procedures to ensure retention of material, as opposed to lengthy explanations. In a classroom setting, with a small class size and the instructor leading the work, these exercises require roughly 40 hours to complete. If you work through this book on your own, you will be slowed somewhat by the lack of instructor prompting, but on the other hand, you will not have to adhere to the pace of others. VIZ is a feature-rich, powerful program that takes some time to learn. You need to commit at least ninety minutes a day to it to get up to speed quickly. At that pace, it will likely take close to a month to get through this book. While that may seem like a long time, consider that grossly optimistic time schedules in book titles and training advertisements ("Learn ____ In A Day") have created unrealistic expectations of learning durations. If you are new to visualization software, VIZ is unlike any program you are likely to have used. Approach the study of it as you would a new language. But on the bright side, it is an extremely fun program to learn and use, and you'll get to bed late many a night, as you get absorbed in the details of the scenes you will soon be creating. This is a highly addictive program.

Level of Student

This book is for the beginning to intermediate user of 3D Studio VIZ. If you have never seen VIZ, that's fine. You will start with the basics, and each step will be explicitly described and illustrated. If you have some experience with VIZ, you will not be bored with this book. The tutorials start at a basic level and proceed to a fairly advanced level. Most users of VIZ will find challenging topics in the exercises.

While no prior knowledge of VIZ or of CAD programs is needed to use this book, learners should be familiar with computers in general, and be comfortable using software of some kind. If you have limited experience with computers and are not entirely comfortable in the computer environment, you will likely be overwhelmed by VIZ.

Hardware and Software Requirements

3D Studio VIZ is a resource-hungry application. Your computer needs at least a Pentium2 processor, and at least 128 MB RAM. A video card with 4 MB of video RAM will suffice (barely). The tutorials should proceed smoothly with these minimums met, but fortunately most computers today have Pentium3 processors and video cards with at least 16 MB video RAM. With these specifications and 256 MB RAM, you will spend much less time waiting for scenes to render during the exercises.

VIZ can run under Windows 98, Windows NT4, or Windows 2000. It will not run under Windows 95. While there are no major issues with using VIZ under Windows 98, the program will run more reliably under Windows NT. Windows display resolution needs to be at least 1024 x 768, and displaying 24-bit color. A display resolution of 1152 x 864 or higher will provide a more comfortable working environment in VIZ.

VIZ takes advantage of a wheel mouse if you have one, but it is not necessary. The wheel mouse provides handy pan and zoom functions. A sound card is not needed for these tutorials.

The tutorials make use of several bonus bitmap collections and material libraries that are optional choices when installing VIZ. If VIZ was installed on your computer using the Typical installation choice, then these extra bitmaps and libraries are not installed on your computer. If you are not certain that installation type Custom was used during the install, and that all the bonus bitmaps and libraries were installed, you should revisit the installation procedure before trying to use this book. You do not need to reinstall the program files or hardware lock drivers or any portion of the application other than the bitmaps and material libraries (matlibs) in the Samples and the Bonus sections. Below is a screen capture showing which items need to be checked.

Transferring Files from the CD

The Viztutorials folder on the CD that accompanies the book contains all the files needed to complete the tutorials, and this folder needs to be copied to your C: drive. While you might prefer to work directly from the CD, the copying of files to the C: drive ensures that references in the book to locations of files are correct and consistent. It also ensures that VIZ will find all the bitmaps that are incorporated into materials used in the scenes. The files to be copied are not large- they take only a few megabytes of disk space. In Windows Explorer, simply drag the Viztutorials folder from the CD and drop it onto your C: drive. When you copy files from a CD, those files are given "Read-only" properties, meaning that if you open one of the files from the C:\Viztutorials folder, edit it, and attempt to save under the same name, you will be unable to do so. The tutorials all prompt you to save under different names, so the Read-only property should not be an issue. If for some reason you want to remove the Read-only property of any of the files, highlight the file name in Windows Explorer, right-click over the name, and choose Properties from the menu. In the Properties dialog box you will find the Read-only check box.

If you have verified that the Bonus and Sample bitmaps are installed, and you have copied the Viztutorials folder from the CD to your C: drive, then you are ready to start.

Bonus Resources on the CD

There are two folders on the CD. The Viztutorials folder contains the support files for the exercises. The Vizresources folder contains bitmaps, models, and plug-ins for use with VIZ, supplied by several manufacturers and distributors. To browse the content of the Vizresources folder, drag the file vizwares.htm from the Vizresources folder and drop it on your Web browser. In addition to the free models, bitmaps, and plug-ins, you will find links to the Web sites of these manufacturers and distributors, so that if you have Web access you can follow those links to see what other VIZ resources are available.

Acknowledgments

Thanks to all the professionals who provided reviewing and editing assistance with this book, especially Jason Busby of The Renaissance Center, Dickson, TN for his technical editing, Pamela Lamb of Watervliet, New York for her copy editing, and reviewers Peter R. Lukasiewicz, sage Engineering Associates, LLP, Albany, NY, and R. Thomas Trusty, Ivy Tech State College, Indianapolis, IN.

Thanks to the staff at Delmar for their responsiveness during a challenging production schedule, and for being patient as I learned the ropes.

Thanks to MicroCAD Training and Consulting of Watertown, MA for coordinating the classes in which these tutorials were tested and tuned.

Dedication

Thanks to my parents, William and Harriet, for all the years of love, support, and patience beyond the call of duty. Thanks to Alex, for being a great friend and for inspiring me to my second career in hockey.

VIZ Process Overview

This first chapter is intended to get you comfortable with the environment of VIZ, and give you an understanding of the overall process of building a scene. You will open a partially completed scene, create a few additional objects, make modifications to some existing objects, learn to navigate around the model, add materials to the objects in the scene, set up a camera, and render a still image like the one shown here.

The fact that you have decided to immerse yourself in a book of VIZ tutorials probably means you've discovered that VIZ is not the type of program that you can teach yourself by playing around with it for a while. The VIZ interface is feature-rich; there are button types you have probably never used, unique mousing maneuvers, fields that roll up into a single button, and so on. VIZ is a more conceptual program than a word-processor or a paint program. While it is important that this book conveys the concepts upon which the functions of VIZ are built, it is not necessary to bore you with long explanations of the ideas. As you work through the tutorials and use VIZ on your own, you will start to get an intuitive sense for the logic of the program, and you will come to appreciate the program's complexity, because in the case of VIZ, complexity equals power and flexibility. If after working through a series of steps in this chapter, you are not sure you could reproduce those steps on your own, don't worry– that is not the intent of the chapter. You will have plenty of opportunity to repeat and reinforce skills in later chapters.

Customizing the Interface

The first time you start VIZ, it should look like this:

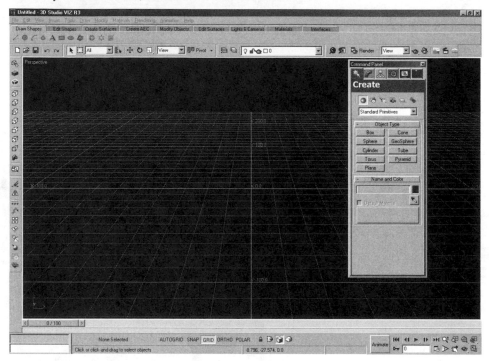

The VIZ interface is easily customized, often by right-clicking the mouse over areas to access menus. Begin by docking the six panels:

1. Place the cursor over the blue title area of the floating window labeled Command Panel, and right-click

2. Choose Dock / Right

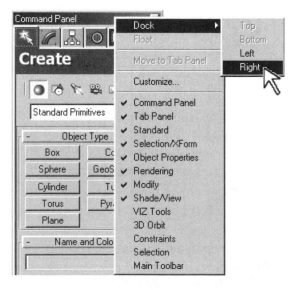

Display an additional toolbar:

3. Right-click over an empty toolbar area at the lower-left of the interface, and in the menu, choose Constraints

4. Drag the Constraints toolbar by its blue titlebar and dock it vertically at the lower left:

If your display is set to 1024 x 768, there may not be room to dock the toolbar. In that case, just close it.

Add a new button to a toolbar:

5. Right-click over an empty area of a toolbar again, and choose Customize

6. In the Customize User Interface dialog box, select Commands

7. In the list of commands, scroll to and select Shade Selected Mode

8. Select the Image button

9. Open the Group drop-down list and select ACAD_Shade

10. Select the rightmost button

11. Drag the button and drop it onto the Shade / View toolbar (upper left)

12. Close the dialog box

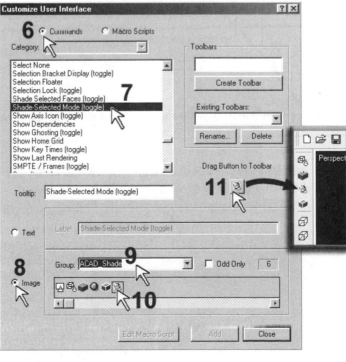

3

The behavior of the Customize User Interface dialog box is a bit unusual. The customization of most programs is set up so that if you select a certain command, a button with an appropriate icon automatically appears. Each icon is programmed to associate with a certain command. This is not the case in VIZ. You can choose any button from any library to associate with a command.

Across the top of the interface is a row of Tab Panels:

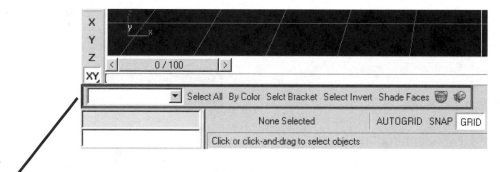

These are just toolbars with tabs on them. Any toolbar can be converted to a Tab Panel, and any Tab Panel can be converted to a toolbar. Call up another toolbar and convert it to a Tab Panel:

1. Right-click over an empty toolbar area and display the Selection toolbar

The selection toolbar will dock itself under the viewports, at the left side.

2. Place the very tip of the cursor at the very edge of the Selection toolbar so that a couple sheets of paper show on the cursor, then right-click

3. From the right-click menu, choose Move to Tab Panel

4

The Selection toolbar now exists as a Tab Panel.

The black background for the viewports was likely chosen to resemble AutoCAD. Maybe you like a black background, or maybe it is a bit severe. Try setting it to gray:

1. From the menus, choose Tools / Options

2. Select the Colors tab

3. In the Main UI group, open the drop-down list, scroll near the bottom, and select Viewport Background

4. Click the color swatch, and in the Color Selector, set the background to Hue = 0, Saturation = 0, Value = 180

5. Choose OK to close the Color Selector, choose OK to close Options

If you decide you like the black background better, feel free to change it back. Note that this color is just for the background of the viewports - it has nothing to do with the background of a rendering.

These changes that you've made will be saved to a file called Vizstart.cui when you close VIZ, and this is how the interface will look next time you start.

6. Close VIZ (File / Exit). If a dialog box appears stating "The design has been modified. Do you want to save your changes?", choose No

7. Restart VIZ

Areas of the Interface

There are eight general areas of the interface:

1. Menu Bar

2. Tab Panels

3. Toolbars

4. Viewport(s)

5. Command Panels

6. Status Bar

7. Time Controls

8. View Navigation Controls

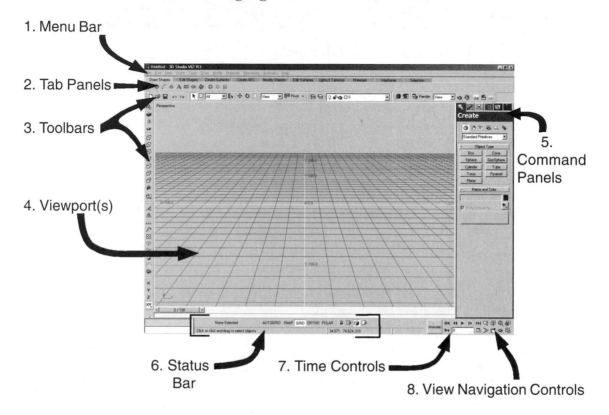

In the workflow, you might move among the areas like this:
- In View Navigation controls, get a good view of the work area.
- In the Command Panels, go to the Create panel and create an object.
- Switch to the Modify panel and alter the object somehow.
- From the Toolbars, choose a tool to relocate or reorient the object.
- Back to View Navigation to center the object in the view and inspect it in 3D.

You won't need the Menu Bar often because most of the items in the menus are duplicated in toolbars and panels. The Tab Panels are completely redundant; they duplicate some functions of the Create panel, the Modify panel, and the Material Editor. To streamline your learning and to simplify the exercises, the Tab Panels are largely ignored throughout this book, but you might try working with them a bit and see if they suit you. One of the handiest things to do with the Tab Panels is to create your own custom toolbar and convert it to a Tab Panel.

Units

It is important to be aware that units are dealt with in two places in VIZ. There is a setting for System Unit - the internal unit of measure that VIZ uses to calculate the size and location of everything in a scene. By default, this unit is the inch. If you only model things as big as a building or machine, and not entire cities or transoceanic tunnels, then you will never need to change this setting. The other unit setting is for Display Units - how the VIZ interface shows you sizes and distances, and how you indicate sizes and distances to the program. The Display Unit determines what you mean when you type the number 1 - is it one inch, one foot, one meter, and so forth.

System Unit

Take a look at where this setting is found.

1. From the menus, choose Tools / Options

2. Choose the General tab

The System Unit Scale is shown as the inch, and you should leave it that way, unless you are building something very large, or very small.

Below the System Unit display is a sliding scale labeled Origin. This is just informational. Accuracy in VIZ does degrade if your scene has objects very far from the 0,0 point (the Origin), and this scale tells you what accuracy you can expect as you build further away from the origin.

3. Choose OK to close the Options dialog box

Display Units

When modeling a machine part you may want to work in millimeters. For a model of a house, you probably want feet, and for sitework, maybe decimal feet or meters. This choice is made in Drafting Settings.

1. From the menus, choose Tools / Drafting Settings / Units Setup

2. Set the display units to US Standard, leave the drop-down list set to Feet w/Decimal Inches, set the Default Units to Feet, and choose OK

3. Type W to see four viewports

Save a Startup File

To avoid having to set the display units every time you start a new VIZ scene, save this empty scene as the default start-up file. When VIZ starts it looks for a file in the \3dsviz\Scenes folder called Vizstart.max. If it finds a Vizstart.max, VIZ opens an unnamed, unsaved copy of that file.

1. From the menus, choose File / Save

2. Verify that the Save In: folder is the Scenes folder (if it is not, browse to the \3dsviz\Scenes folder)

3. Name the file Vizstart (no need to type the .max on the end, VIZ will add that for you)

4. Choose Save

5. From the menus, choose File / Reset to open a new, unsaved scene using the settings you just saved into the Vizstart.max. At the prompt "Do you really want to reset?", choose Yes

Open a Scene

1. From the Toolbars, choose the Open Design button

2. In the Open File dialog box, browse to the C:\Viztutorials\Chapter1 folder, highlight StillLife01.max, and choose Open

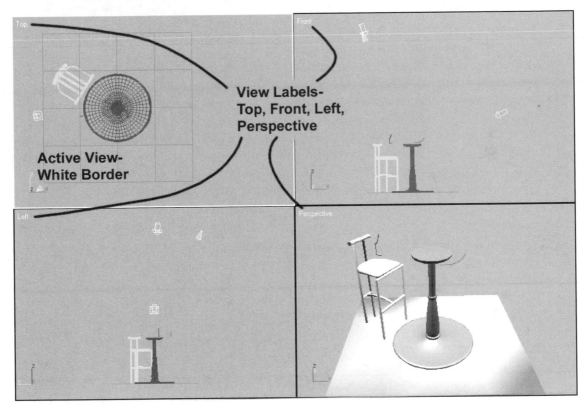

View Navigation

The scene was last saved with four viewports displayed. Each view is labeled at its upper-left: "Top, Front, Left, Perspective". The Top view is active – it has a white border. Switch active views:

1. Position the cursor anywhere over the Perspective view, and *right-click*

> ⚠ It is essential that you switch views with a right click, and not a left-click. In most situations the active command in VIZ is the Move tool. If you develop the bad habit of changing active views with a left-click, and your cursor happens to be over an

9

object as you switch views, you will probably nudge the object, and you may not even know you've done so.

2. From the View Navigation tools, at the lower-right of the interface, choose the Zoom tool

3. Position the cursor near the center of the Perspective view, and drag up and down to zoom in and out. Dragging side to side doesn't do anything. Zoom back out to approximately the zoom you had before

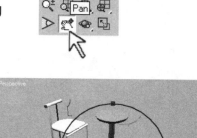

4. Try the Pan tool; it is self-explanatory. After trying it, pan the scene back to about where it was

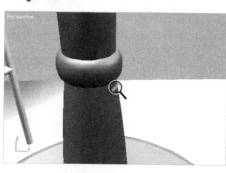

5. Rotate about the scene: Choose the Arc-Rotate Selected tool. A green circle appears in the Perspective view. Place the cursor anywhere *inside* the circle (if you start outside, the view tilts) and drag in all directions. Arc-Rotate back to about where you began, and release the mouse. Right-click to leave Arc-Rotate mode

There are lights in the scene, placed several feet away from the objects. Use Zoom Extents to see them in the Perspective view:

6. Choose the Zoom Extents tool (or Zoom Extents Selected – either will work)

Zoom Extents is more useful if you control what objects are considered in determining the extents of the scene. You will not work with the lights in

10

this exercise, and it would be handy if you could instruct Zoom Extents to ignore the lights, and just zoom to show all the geometry. Ignore Extents does this.

7. Make sure the Select tool, found near the left end of the toolbar above the views, is active (green)

8. Make the Front view active (with a right-click!)

9. Drag a selection around the three lights (they will turn white when selected). Be careful not to select anything but the lights- if the selection window touches anything else, it will select it. If you do accidentally select other objects, just drag the selection window again to make a new selection

10. With the lights selected, choose, from the menus, Modify / Properties

11. In the Object Properties dialog box, check Ignore Extents, then choose OK

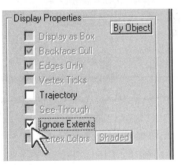

12. Place the cursor over any empty area of the Front view and click, to deselect everything

13. Make the Perspective view active

14. Zoom Extents (or Zoom Extents Selected) again

This time the objects should fill the Perspective view, with the lights being disregarded.

15. Use Zoom Extents All to zoom tight to the geometry in all four views

16. Use the Min/Max Toggle tool to have the Perspective view fill the viewports area

Set the Perspective view to a wireframe display:

17. Right-click over the view label (the word Perspective)

18. From the menu, choose Wireframe

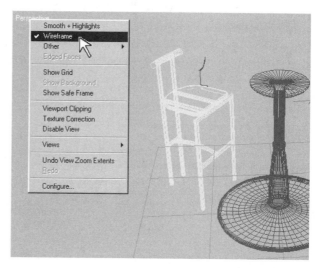

There is a better way to do this, using a keyboard shortcut:

19. Type the F3 function key a couple times to toggle between shaded and wireframe displays, then leave the view in wireframe display

The Perspective view makes a nice-looking working environment, but it can be difficult to work in perspective, and view navigation is a bit easier in an axonometric view. The VIZ name for an axon view is a User view.

20. Type the letter U, then Zoom Extents again

The perspective is gone.

Look at the View Navigation tools - they've changed. While in perspective there is a Field-of-View tool. In axon that changes to Region Zoom - a familiar window zoom, and a much more useful function.

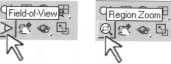

The Create Panel

Time to make something. The table base needs a round top, which you will make from a category of objects called Primitives. Primitives are variations of basic geometric forms like boxes, cylinders, spheres, and prisms.

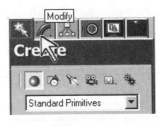

1. Place the cursor on the tab of the Create panel, so the tooltip shows, then move across all six panel tabs and read the tooltips

You will spend the most time, by far, in the **Create** and the **Modify** panels. The **Hierarchy** panel is for describing how moving parts relate to each other. You would use this panel extensively if you were animating a robotic assembly device. If you are an architect, you might find two common uses for the **Motion** panel; it is where the motion of the Sun across the sky is controlled, and it is where you would fine-tune the motion of a camera through a building. The **Display** panel is for showing and hiding objects, and for locking objects so they cannot be changed. Most of the functions of the Display panel are duplicated in a small floating window called the Display Floater, and you will probably find that floater more efficient than switching to the Display panel. The **Utilities** panel holds a collection of useful routines that do a wide variety of specialized tasks. You will most commonly go to the Utilities panel to open the Asset Browser, a drag-and-drop interface for filling your scene with models and materials.

Just below the word Create is a row of six buttons. These are Categories of objects: Geometry, Shapes, Lights, Cameras, Helpers, and Systems.

Just below the Category buttons is the Subcategory drop-down list. By the end of this book, you will have explored all but one of these subcategories.

2. With the Geometry category chosen, open the subcategory drop down list and choose Extended Primitives

13

3. Choose ChamferCyl

Notice that the active tool is lit green. In VIZ the active tool is "modal". You are now in ChamferCylinder-making mode, and after creating one, the button will still be lit green, waiting for you to create another one. You can turn off a modal tool by either right-clicking the mouse, or typing the Esc key. Several of the View Navigation tools are modal.

Scrolling Panels and Rollouts

In many panels there are more tools than can fit on the screen at once. This is dealt with in two ways: scrolling panels, and rollouts.

Scrolling panels: when a panel's contents disappear off one end, place the cursor in a blank area of the panel, wait for a white hand to appear, then drag the panel up and down to see its entirety.

Rollouts: These are wide buttons (bars) with either a plus or a minus sign at their left end. A plus means there is content rolled up into the bar. A minus means you can roll unneeded content up into the bar.

4. Click on the Keyboard Entry rollout to open it

5. Place the cursor in a blank part of the Create panel, and when the cursor shows a white hand, click and drag upwards to see the rest of the ChamferCyl parameters

Here you can see the advantage of having your screen resolution set high (1280 x 1024) and having a big monitor - you'll do much less scrolling.

If you right-click when you see the panel-scroll cursor (the white hand), you will see a menu that you may find useful on really huge panels with several rollouts. You can very quickly jump to any area of the panel by selecting the name of a rollout. For example, to go to the top of the panel, you would select Object Type. To go to the middle of the panel you might select Creation Method.

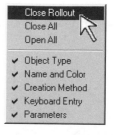

Spinners

As you can see, creating objects involves setting values in numerous numeric fields called Spinners. Spinners are used extensively throughout VIZ. It is important to develop good habits for entering values in spinners. The fast, efficient way is to use the number pad and the up and down arrows on the keyboard. Click in the first spinner to highlight the contents, type the value using the number pad (make sure Num Lock is on), then type the down arrow to jump to the next spinner and highlight its numeric value. You can quickly enter values in several spinners without touching the mouse.

6. In the Keyboard Entry rollout, click in the spinner labeled Z: to highlight the value, then type 3'2" (this will create the tabletop at the correct height on top of the table base)

7. Type the Down arrow to the left of the number pad to jump to the Radius spinner, enter a radius of 2'

8. Arrow down to the Height spinner, type 1"

9. Arrow down to the Fillet spinner, type .25"

Fillet is the easing of the edges of the cylinder, creating a radiused transition between sides and top or bottom.

Note that there is a second set of spinners for Radius, Height, and Fillet, located in the Parameters rollout. Ignore these for now - the spinners in the Keyboard Entry rollout take precedence.

10. In the Parameters rollout, click to highlight the Sides spinner; type 60

11. Put a check in the box labeled Smooth

12. Choose the Create button

The tabletop should appear in the scene, located correctly atop the table base.

There are a couple more facts about spinners you need to know. The arrows at the right of a spinner work in two ways. You can click either arrow to nudge the numeric value up or down. You can click-hold on either arrow (it doesn't matter which) and drag up and down to interactively change the numeric value. The object will respond simultaneously in the views.

When typing numeric values, you don't need to type every foot and inch symbol. You have set VIZ up with a default Display Unit of the foot. This doesn't mean that the scale of the scene is based on the foot (by default the scene unit is the inch). It means that if you type 5, VIZ interprets that to mean 5 feet. So to set a radius of 2 feet in a spinner, just type 2, and forget about the foot symbol. There's also a common situation in which there is no need to type the inch symbol. If the numeric value is 6 feet 4 inches, just type 6′4. VIZ knows you mean 6 feet 4 inches. These are small details, but little things add up.

Wheel Mouse

If you have a wheel mouse, take a moment to explore the Pan and Zoom functions it offers:

You can pan a view by dragging with the wheel. When you click-hold the wheel, the cursor becomes a Pan symbol.

Roll the wheel to zoom in and out.

There is a setting in the Options called Zoom About Mouse Point, and it is active by default. When doing a simple interactive zoom, it matters where you place the cursor as you begin the zoom. If you start the zoom with the cursor in the middle of the view, the view stays centered as it is. But if you start the zoom with the cursor near a corner, for example, the zoom action will be weighted toward that corner, and you will be panning at the same time you zoom. So before you start to drag a zoom, give a moment's thought to what you want to see, and therefore where you should place the cursor

The wheel mouse Pan/Zoom seems to not always respond. If you roll the wheel to zoom, and nothing happens, try a wheel mouse Pan first, then switch to Zoom.

Modifiers

Objects in VIZ carry a history. Things start out being defined by their Creation Parameters, which in the case of the ChamferCyl you just made are the Radius, Height, Fillet, and so on. At any time, you can access these Creation Parameters and edit them to alter the original object. You can also alter the object by adding Modifiers to it. Modifiers are packets of programming that have names like Extrude, Bend, Twist, Taper, Noise, and Slice. The Modifiers sit on top of the Creation Parameters in the object's history, and VIZ processes the history, starting with the Creation Parameters and working up through however many Modifiers you've piled on, to finally arrive at the appearance of the object in the views and in the renderings.

There are a couple 2D shapes located on the tabletop. Add a Modifier to one of them to make a 3D wine glass.

1. From the toolbar above the viewport, choose the tool labeled Select By Name

2. In the Select By Name dialog box, highlight Wineglass01, and choose Select

3. Switch to the Modify panel

Near the top of the Modify panel are ten buttons with names like Slice, Cap Holes, Edit Mesh, and UVW Map. These are ten of several dozen possible Modifiers that might be added to an object (the entire list is accessed via the button labeled More).

4. Click once on the Lathe modifier

You need a close-up view of the object. There is a zoom tool called Zoom Extents Selected, which zooms the view tight to the selected object. This tool is nested under another zoom tool, in a type of button called a Flyout button.

5. In the View Navigation tools, click-hold on the Zoom Extents tool. The flyout button expands, and you see two buttons contained in the flyout – a gray version and a white version. Move the cursor over the white version (Zoom Extents Selected), and release

The 3D object is interesting, but it sure doesn't look like a wine glass. It has lathed itself about the centroidal axis of the 2D shape, and it should lathe about the left-hand side of the shape.

6. Near the bottom of the Modify panel, in the group labeled Align, are three buttons labeled Min, Center, Max. Choose the one labeled Min

Now it looks recognizable.

7. Set the Segments spinner to 20

This gives more "wedges" to the object, so it's better defined.

8. Check the Generate Mapping Coordinates checkbox

Mapping Coordinates are instructions for how a bitmap should wrap itself onto the object. Mapping Coordinates would allow you to add an etched design on this glass.

9. Uncheck the Generate Material IDs checkbox

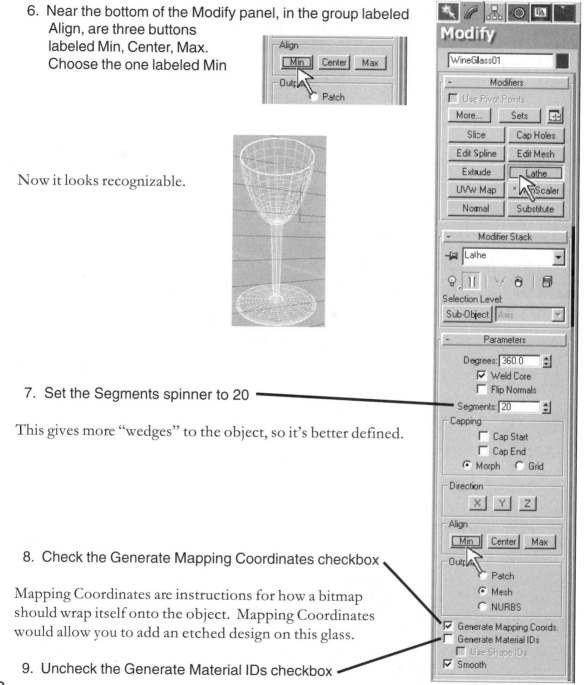

18

Creating Shapes

To create objects, you can add Modifiers to 3D objects or to 2D shapes, as you just did with the wine glass. There are a few items that appear in the views in 3D right away, such as Walls, Doors Windows, and Stairs. But most objects begin as 2D shapes and have Modifiers added that make them 3D. So you'll spend considerable time drawing and editing 2D shapes. Draw a many-pointed Star, then modify it into a fancy dish:

1. Zoom and Pan to see all of the tabletop

2. Switch to the Create panel

3. Choose the second category button, Shapes

4. Choose the Star tool

In making the ChamferCyl for the tabletop, you typed all the parameters into spinners, then clicked Create to make the complete and correct object. More often, you'll just drag out objects interactively in the views, not worrying about their parameters, and then change the parameters to correct the object after it is created.

Autogrid

Autogrid mode allows you to create objects on whatever surface you want. Without Autogrid, the Star you are about to drag out would be located on the floor. Use Autogrid to drag the Star out on the tabletop:

5. At the Status Bar at the bottom of the interface, choose Autogrid

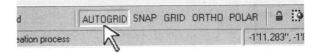

An axis tripod appears at the cursor to let you know you are in Autogrid mode. The cursor itself also changes; the thin cross is a Pick cursor, which means you are in a command or a mode that involves clicking on the surface of an object.

19

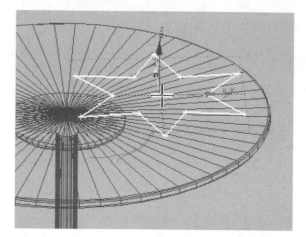

6. With Autogrid active, position the cursor anywhere on the tabletop, click-hold, and drag out a Star shape. A complete Star appears, but when you release the mouse, you're not done. Release, then move the cursor up and down to change the size of the points. Click to set the points to any size

7. As soon as the Star is made, edit its parameters in the Create panel. Remember to use the number pad and arrow keys to change the spinner values

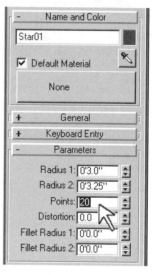

   Set Radius 1 to 3″

   Set Radius 2 to 3.25″

   Set Points to 20, then type the Enter key to see the points update

The shape doesn't look like a star anymore; now it is a fluted circle.

Add a Modifier

There is a curvy profile shape visible on the tabletop. That shape is the half-silhouette of a dish that will hold fruit. There is a Modifier that creates a 3D object from two 2D shapes: the footprint of an object (the Star) and a silhouette of the object.

8. Turn off Autogrid

9. Switch to the Modify panel

10. The Modifier you need is called Bevel
    Profile. It is not on the ten buttons show-
    ing, so choose the More button to bring up
    the complete list of available Modifiers

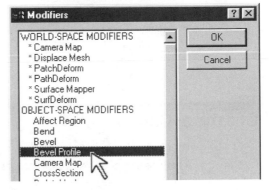

11. Select the Bevel Profile modifier, and
    choose OK

12. In the Modify panel,
    choose the Pick Profile
    button – it turns green

13. Position the cursor over
    the curvy profile shape,
    and when the cursor
    appears as a small
    white selection cross,
    click

How does VIZ know you
want to pick the curve, and not the tabletop? A Bevel Profile path cannot be a 3D
object like the tabletop; it must be a 2D shape like the curve, so VIZ knows to pick the
curve.

At this point, you should
see a fluted dish:

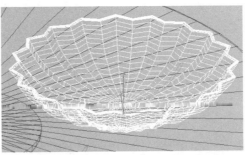

Save the Scene

1. From the menus, choose File / Save As

2. In the Save File As dialog box, browse to the C:\VIZtutorials\Chapter1 folder, name the file StillLife02.max, and choose Save

Create a Camera

You are done modeling objects, and you are about to start arranging and copying objects to make a composition, so you need to establish the camera view for the rendering.

1. To restore four views, type the letter W

You can switch from one view to four with the Min/Max Toggle tool in View Navigation, but typing W is probably easier.

2. Make the Top view active

3. Zoom out a bit in the Top view, as shown here:

4. Switch to the Create panel

5. In the categories buttons, choose Cameras

6. Under Object Type, choose Target

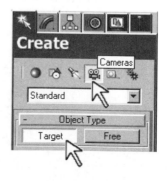

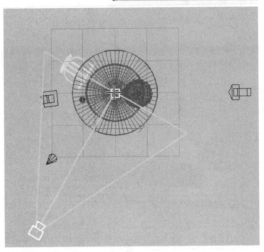

7. From the Stock Lenses buttons, choose 35mm

8. Drag a Camera in the Top view, starting off the southwest corner of the scene, and releasing the mouse over the center of the table, as shown here:

9. Right-click to leave Camera-making mode

10. Make the User view active

11. Type the letter C to change the User view to a Camera view

12. Zoom Extents All to see the entire Camera in all views

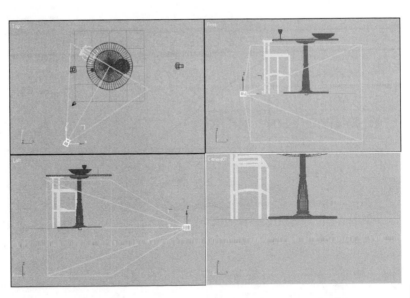

Look at the View Navigation tools; they've changed. View Navigation tools are context-sensitive. They will change their tools as you switch to different types of views. To change the view through a Camera, you can either relocate the Camera and its Target in the Top, Front, and Left views, or you can work through the Camera view itself, using the View Navigation tools to compose the view.

13. From the View Navigation tools, choose Truck Camera

To Truck a Camera is what a CAD program would call a pan - the Camera and the Target move in unison.

14. Dragging in the Camera view, center the objects in the view, as shown here:

15. From the View Navigation tools, choose Orbit Camera

16. Drag downward in the Camera view to get a view looking down on the table:

Canceling Commands

This is a good time to practice canceling a command in progress. It is done with a mouse-click combination, and it works for any command/action.

1. Begin a new Orbit Camera action. Instead of releasing to set a new view, keep the left mouse button held, while you right-click. The Orbit action will cancel. Release both mouse buttons

2. Right-click again to leave Orbit Camera mode

Display Panel

There are a number of other objects modeled and present in the scene, but their display is turned off. Use the Display panel to show them.

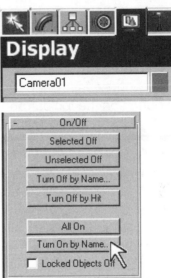

1. Switch to the Display panel

2. Choose the Turn On by Name button

3. Highlight these items in the list:
   Apples, Bananas, Candlestick01,
   Grape Bunch, Lemons, Wine Bottle

The easiest way to highlight sequential items in a VIZ list is to click-hold on the first item and drag, to highlight the rest.

4. Choose On

With the additional objects displayed, you probably need to see a bit more of the scene.

5. Use Dolly Camera and Truck Camera to move the Camera back some and re-center the scene in the view, as shown here:
   Leave a bit of room above the candle

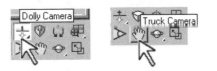

Transforms

The Transforms are Move, Rotate, Scale, and variations on these three. Transforms are the location in space, the orientation, and the size of an object. If you perform Transforms with the Shift key on the keyboard held, you will make duplicates of objects. Use Transforms to position the fruit dish under the fruit, and to make a couple more barstools and wine glasses.

1. Save your work so far. Choose the Save Design button on the toolbar above the views

2. Make the Top view active

3. Type W to maximize the Top view

4. Use Region Zoom to zoom in on the fruit dish and the fruit. To use Region Zoom, place the cursor at one corner of the desired view, click-hold, drag a rectangle representing the view you want, and release

5. Choose the Select by Name tool

6. From the list, highlight Star01, choose Select

7. Just to the right of Select by Name, choose the Select and Move tool

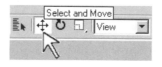

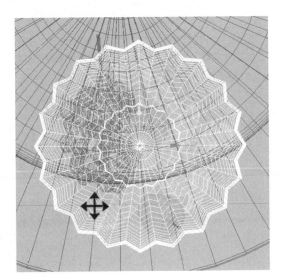

8. Position the cursor over the lower-left of the fruit dish, make sure the cursor appears as a four-way arrow, and drag the dish until it's centered under the fruit, then release

Making Duplicates

Duplicating, or Cloning as it is called in VIZ, is most commonly done by performing a Transform with the shift key held.

1. Use the Select by Name tool to select WineGlass01

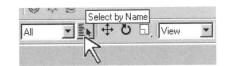

2. Use the Pan tool to pan the Top view to see the area where the wine glass and wine bottle are

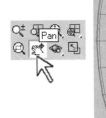

3. Choose the Select and Move tool again

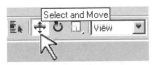

4. Position the cursor over the wine glass, make sure the cursor shows the four-way arrow, hold the Shift key on the keyboard, and drag the wine glass toward the upper-right. Position the new wine glass as shown below, and release the mouse and the Shift key

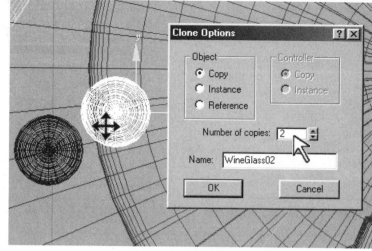

5. In the Clone Options dialog box, set the Number of Copies to 2, then choose OK

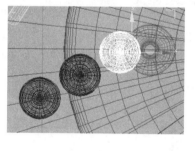

The three wine glasses should look like this:

6. Move the third wine glass (the selected one) so it is directly above the second one:

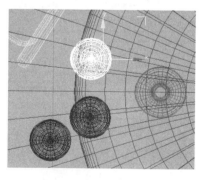

7. Zoom out to see the whole table and the stool, then right-click to leave Zoom mode

8. To select the stool, place the cursor directly on it, and when the cursor shows a small white selection cross, click

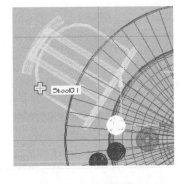

9. Hold the Shift key on the keyboard, and drag a second stool, as shown here:

10. In the Clone Options dialog box, set Number of Copies to 1, then choose OK

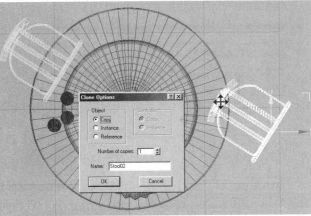

11. Shift-Move the second stool to create a third one:

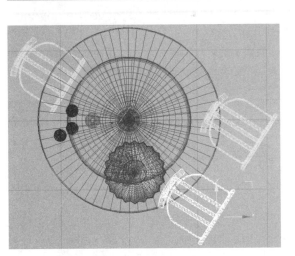

12. Next to the Select and Move tool, choose the Select and Rotate tool

13. Attached to the stool is an axis icon. The X axis is red, the Y axis is green, and the Z axis, which points directly at you, is blue. Position the Rotate cursor over the blue Z axis, and drag the mouse up and down to rotate the stool. Rotate it to a position similar to this:

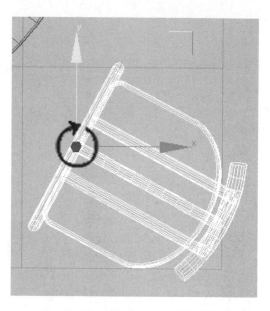

14. Switch to the Select and Move tool and move the stool halfway under the table

15. Type H, and from the list, select Stool02

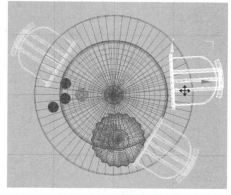

16. Switch to Select and Rotate, and rotate the second stool to face the table

17. Switch back to Move, and move the second stool halfway under the table:

As you're moving the stools, you may discover that if you place the cursor on one of the axes of the axis icons as you begin to drag, you can only drag in one direction. The functions of the axes will be fully explained in the next chapter. For now, just drag on a different part of the object.

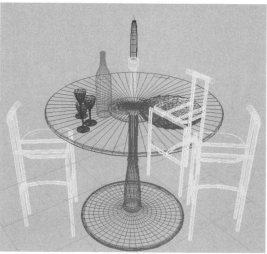

18. When the three stools are correctly positioned, type W to return to 4 views, make the Camera view active, then type W again to maximize the Camera view

19. Save the scene (File / Save)

Rendering

1. There are three green teapots at the upper-right of the interface. Choose the middle of the three, Quick Render

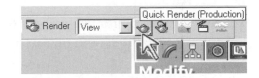

A floating window (called the Virtual Frame Buffer) appears, and the renderer begins processing the geometry of the scene from the top down. Once it reaches the bottom, a second pass calculates a post-processing effect to add a glow from the candle.

Even without proper materials on most of the objects, it is still an evocative image, mainly because it has interesting lighting and shadows. Light, shadow, and reflection do much more for an image than detailed modeling and photorealistic materials do.

Notice the floor in the rendering, versus the floor in the viewports. In the views, it is a small square barely large enough for the table and stools, but in the rendering the floor extends into the darkness. The floor is a Plane object, and one of the parameters of a Plane object is the Render Multiplier, which in this scene is set to 2.2, so that the floor will be rendered 2.2 times the dimensions it appears in the views.

The rendering is comprised of three channels of color. It's interesting to see those channels isolated. Try the three channel enabler buttons at the top of the rendered view, just to see what they do.

Materials

The last thing to do for this scene is to pull some premade materials from a material library and drop them onto objects in the views.

1. Close the Virtual Frame Buffer

2. From the menus, choose Materials / Material/Map Browser

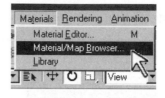

3. Drag the Material/Map Browser to one side of the interface so you can see as much of the scene as possible behind it.

4. Near the upper-left of the browser, select Browse From / Mtl Library

5. Near the lower-left of the Browser, choose File / Open

6. In the Open Material Library browser, browse to the C:\VIZtutorials\Chapter1 folder and open StillLife.mat

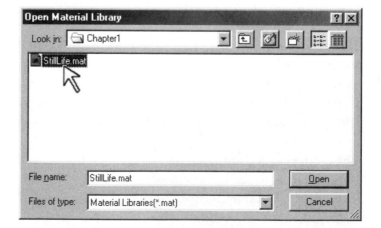

7. Single-click on the Barstool material in the list to see a sample sphere for the material

8. Drag the material from either the name in the list, or from the sample sphere, and drop it onto one of the stools in the view

9. Drag the Barstool material onto the other two stools. As you drag the material the second time, you may see a dialog box asking about duplicate material names. Choose Replace It

10. Highlight the Floor-Parquet material, then drag it to the floor in the view. You will have to position the cursor on the very edge of the floor to drop the material on

11. Highlight the Fruit Dish material and drag it onto the fruit dish. The dish is in a crowded area. To ensure that you are dropping the material onto the correct object, position the cursor over the object and leave it there a moment. A tooltip identifying the object (in this case Star01) will appear. Release the mouse

12. Drag the Table Base material onto the table base, and drag the Table Top material onto the tabletop (ChamferCyl01)

13. To assign the Wine Glass material, first use the Select by Name tool to select all three wine glasses.

When you drag the Wine Glass material onto the selection of glasses, the Assign Material dialog box appears. Choose Assign to Selection

14. Once the materials are assigned (no need to assign the fruit, candlestick, or wine bottle materials – they're already assigned), close the Material / Map Browser

15. Save the scene

16. Choose the leftmost render teapot, labeled Render

17. In the Render Design dialog box, choose 800 x 600 as the Output Size, then choose Render

Summary

You should now have an understanding of the workflow - Create objects, add Modifiers, locate the objects with Transforms, add a Camera, add Lights, build and assign Materials, and Render the scene.

As you continue with the book and explore VIZ on your own, don't let yourself slip into unproductive habits. Change views with a right-click, enter spinner values with the number pad and arrows, and if you have a wheel mouse, get used to using it for simple Pan and Zoom functions. Take advantage of keyboard shortcuts; you've learned that W toggles between 1 and 4 views, and that the F3 function key toggles between wireframe and shaded display modes. Several more shortcuts will be introduced in later chapters.

Before moving on to chapter 2, you should work a bit on your own. Try building the simplified version of a still-life shown here, using just primitives. Drag out a few Cylinders, ChamferCylinders, Tubes, and Toruses, use Transforms to arrange them, drag some materials from libraries onto the geometry, and render.

The objects in this scene are found in the Standard Primitives and Extended Primitives subcategories of the Geometry category. The best approach to building the scene is to create the component parts in a rough layout, and then use Transforms to assemble them correctly. Arrange the pieces by eye- next chapter you will learn tools for assembling them with accuracy.

Take advantage of Autogrid to create things at their proper elevation. Once you've dragged out a ChamferCylinder for the base of the table and set its parameters properly, turn on Autogrid before dragging out the Cylinder that is the stem of the table, and the bottom of the stem will sit on the top of the base. With Autogrid on you can create the glasses and donuts directly on the tabletop, instead of creating them at floor level and moving them up.

At the start of this chapter you created the tabletop using Keyboard Entry- entering in all the proper parameters for the object before choosing the Create button. Rather than using Keyboard Entry for these objects, just create them interactively in the Perspective view, by dragging with the mouse, and then switch to the Modify panel to edit the parameters of each object. Creating each object will involve multiple mouse drag and click actions. When you think you are done creating an object, move the mouse and watch the various views– you may see that you are still interactively changing a dimension, and that one more click is required to complete the object.

Use these parameters for the various objects:

Table Base: ChamferCyl

Parameters
Radius: 1'0.0"
Height: 0'1.5"
Fillet: 0'0.25"
Height Segs: 1
Fillet Segs: 1
Sides: 40
Cap Segs: 1
☑ Smooth

Glasses: Tube

Parameters
Radius 1: 0'2.0"
Radius 2: 0'2.25"
Height: 0'7.0"
Height Segments: 1
Cap Segments: 1
Sides: 16
☑ Smooth

Table Stem: Cylinder

Parameters
Radius: 0'2.0"
Height: 3'0.0"
Height Segments: 1
Cap Segments: 1
Sides: 13
☑ Smooth

Milk: Cylinder (vary heights)

Parameters
Radius: 0'1.8"
Height: 0'6.0"
Height Segments: 1
Cap Segments: 1
Sides: 13
☑ Smooth

Tabletop: ChamferCyl

Parameters
Radius: 1'6.0"
Height: 0'1.5"
Fillet: 0'0.25"
Height Segs: 1
Fillet Segs: 1
Sides: 60
Cap Segs: 1
☑ Smooth
☐ Slice On
Slice From: 0.0
Slice To: 0.0
☐ Generate Mapping Coords.

Donuts: Torus

Parameters
Radius 1: 0'2.0"
Radius 2: 0'0.8"
Rotation: 0.0
Twist: 0.0
Segments: 16
Sides: 9

Materials: All are from library 3dsviz.mat, found in the \3dsviz3\matlibs folder

Table Base and Stem: Metal– Chrome Sunset
Tabletop: Metal– Gold Crinkle
Glasses: Glass– Orange
Milk: Plastic– White
Donuts: Ground– Sand Texture

36

CHAPTER 2

Building a Scene: The Barnyard

In this exercise you will assemble a barnyard scene comprised of both simple objects called Standard Primitives, and more sophisticated parametric objects such as Foliage and Railings. You will become more comfortable with the VIZ interface and the navigation of various views, learn to accurately create, position, and duplicate objects, and get an introduction to lighting and cameras.

Start a New Scene, Verify Units

1. If VIZ is already started, choose, from the menus, File / Reset to open an empty, unsaved scene based on the Vizstart.max file you saved last chapter

2. If only a single Perspective view is showing, type W to switch to four views; Top, Front, Left, and Perspective

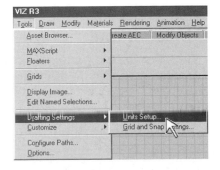

3. From the menus, choose Tools / Drafting Settings / Units Setup. Verify that units are set to Feet w/Decimal Inches and that the Default Units are Feet

Standard Primitives

At the right of the screen
are the six panels, and
the first is the Create panel:

1. Make the Top view active

2. Choose the Create Panel, then
   choose Box

3. Open the Keyboard Entry rollout

4. Set Length to 150'
   Set Width to 150'
   Set Height to -10' (note the minus)

5. Choose the Create
   button

6. Right-click anywhere over the Top
   view to leave Box-making mode

Object Name

1. In the Create panel, click on the name Box01 to highlight it, then type the name Barnyard Base

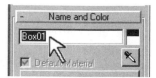

The way you name objects is important. Selecting an object is the first step in accomplishing almost anything in VIZ, and the best way to select an object is often with the Select by Name dialog box. If you are methodical in your naming scheme, objects will be ordered and grouped efficiently in the Select by Name list. Tips on effective naming conventions will be offered throughout the book.

Object Color

1. Click the color swatch to the right of the name to open the Object Color dialog box

2. Set the Palette choice to 3D Studio VIZ palette

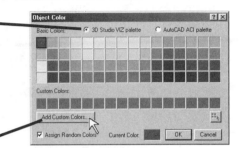

Object Color is the color that an object is displayed in the viewports.

3. Near the bottom of the Object Color dialog box, choose the button labeled Add Custom Colors to open the Color Selector.

4. Position the two open windows so that both are entirely visible

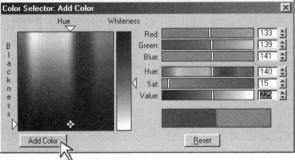

Setting Custom Colors

With these windows open, you will make some new, custom colors that will be appropriate for displaying the various objects in the scene. Don't confuse these display colors with materials. The display colors are being used as a substitute for real materials in this exercise. If no material has been assigned to it, an object will render in its Object Color.

The Color Selector shows three separate sets of sliders for doing one thing; mixing wavelengths of light. The three color models are shown because 3D Studio is used by people from many professions, and different disciplines use different color models. The simplest model is probably HSV, or Hue, Saturation, Value, and it is the model that will be used throughout this book, whenever colors are specified.

5. In the Color Selector, enter values in the HSV spinners of:

Hue = 140
Saturation = 15
Value = 225

Remember to use the number pad and arrows to set these spinners quickly

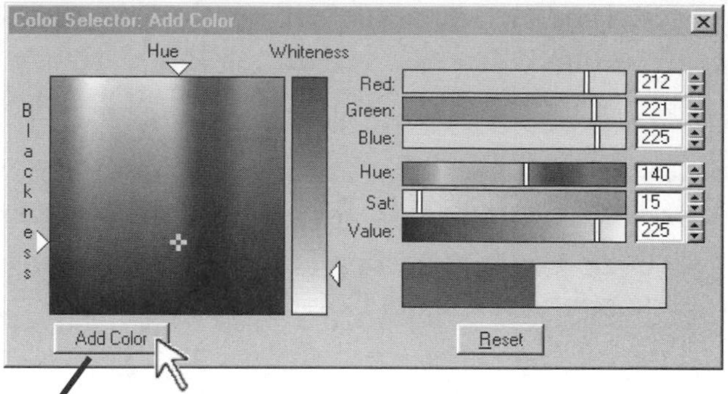

Color Selector: Add Color

Hue   Whiteness

B
l
a
c
k
n
e
s
s

Red: 212
Green: 221
Blue: 225
Hue: 140
Sat: 15
Value: 225

Add Color   Reset

6. Choose the button labeled Add Color, and the silver should appear in the first Custom Colors swatch

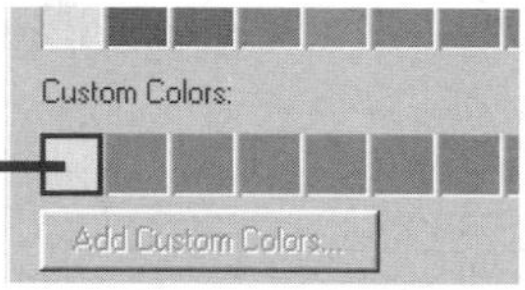

Custom Colors:

Add Custom Colors...

7. Make the 2nd Custom Color swatch active

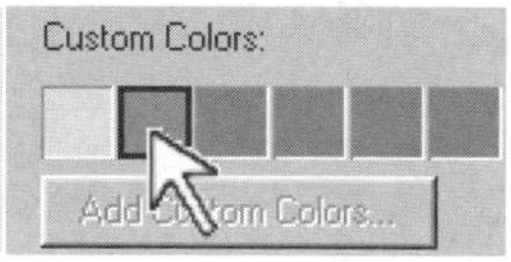

Custom Colors:

Add Custom Colors...

8. Create another Custom Color, with Hue = 140, Sat = 15, Value = 130.

9. Continue creating Custom Colors, according to the figure at right:

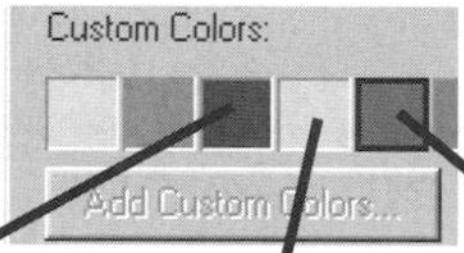

Custom Colors:

Add Custom Colors...

Hue = 0   Hue = 40   Hue = 10
Sat = 0   Sat = 50   Sat = 230
Value = 40   Value = 220   Value = 150

10. Close the Color Selector

11. Before closing the Object Color window, select, from the sixty-four standard VIZ colors, a brownish color for the barnyard base, then choose OK

Zoom Extents All

1. From the eight view navigation tools in the lower-right of the screen, choose Zoom Extents All: Zoom Extents All is found in a flyout button. It is the version of that button showing a gray magnifying glass (a bit hard to see, but it is the upper one)

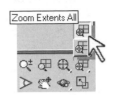

Toggle Grid Display

1. Make each view active and select the Grid button at the Status Line to turn off the grid in each view

Save the Scene

1. Choose File / Save, save the scene into the C:\Viztutorials\Chapter1 folder, naming it Barnyard.max

Creating Primitives

You will work in the Create panel to make parts for the barnyard scene. Don't worry about precise placement of the objects; you will reposition them later. Just create them anywhere near the locations shown in the screen capture images.

You will not use keyboard entry for these items; you'll just drag them out with the mouse, and adjust their parameters after you make each piece.

Creation techniques vary slightly between items, but you begin creating all of them by click-holding, then dragging to see and set the first two dimensions of the object. After releasing the mouse, move it around a bit with no buttons held. With most of the primitives you will find that you are changing another dimension of it (usually the height). Click to set that other dimension. Move the mouse again, and see if the object responds; if it does, you have another dimension to set with a mouse click.

Once you have dragged out and set every dimension of the object, immediately address its parameters in the Create panel.

Just after an object has been created, its parameters are available for editing in the Create panel. If you deselect the object or issue another command, the parameters leave the Create panel, and get moved to the Modify panel. If an inadvertent mouse click causes an object to become deselected before you have a chance to adjust its parameters,

make sure the Select tool is active (it is near the left end of the toolbar above the views),

position the cursor over the object to be selected, and when the cursor becomes a small white cross, click to make the selection.

Then adjust the object's parameters in the Modify panel.

Create Primitives in Top View

As you create these various objects in the Top view, do not worry about their exact placement. Just position them initially in roughly the configuration shown here:

Later, you will use Transform tools to position and assemble things correctly.

1. Drag out a Box for the walls of the barn, then give it the parameters shown here:

Length:	60'0.0"
Width:	30'0.0"
Height:	20'0.0"
Length Segs:	1
Width Segs:	1
Height Segs:	1

2. Name the box Barn. Assign it Custom Color #5, the red

3. Make a Cylinder for a silo

Note that Smooth is not checked. The silo should appear to be made of vertical metal panels; the facets of the panels should show.

Radius:	15'0.0"
Height:	60'0.0"
Height Segments:	1
Cap Segments:	1
Sides:	16
☐ Smooth	
☐ Slice On	

4. Name the cylinder Silo, and assign it Custom Color #1, the silver

42

Radius: 15'0.0"
Segments: 16
☐ Smooth
Hemisphere: 0.5
⦿ Chop ○ Squash
☐ Base To Pivot
☐ Generate Mapping Coords.

[Sphere]

5. Make a Sphere for the top of the silo:
   Note the Hemisphere setting of **.5**, and
   note that Smooth is not checked

6. Name the (hemi)sphere Silo Cap, and
   assign it Custom Color #1, the silver

Radius 1: 20'0.0"
Radius 2: 2'0.0"
Height: 20'0.0"
Height Segments: 5
Cap Segments: 1
Sides: 16
☑ Smooth
☐ Slice On
Slice From: 0.0
Slice To: 0.0

[Cone]

7. Make a Cone for a pile of grain:

8. Name the cone Grain Pile, and assign
   it Custom Color #4, the yellow

Create Primitives in Front View

1. Switch to the Front view, Zoom Extents
   (not Zoom Extents Selected)

Zoom Extents

2. Create a few more parts, arranged
   roughly as shown here:

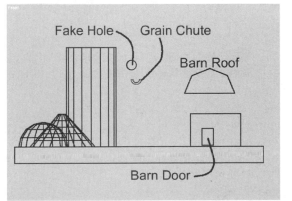

Fake Hole   Grain Chute
                    Barn Roof
                    Barn Door

Cylinder

3. Make a Cylinder for the barn roof:
   Note that Slice is used here, with a
   Slice From value of 180

4. Name the cylinder Barn Roof, and
   assign it Custom Color #2, the gray

Radius:	15'0.0"
Height:	60'0.0"
Height Segments:	1
Cap Segments:	1
Sides:	5
☐ Smooth	
☑ Slice On	
Slice From:	180.0
Slice To:	0.0

Tube

5. Make a Tube for a grain chute:
   Note that Slice is used, with a Slice
   From value of 270, and Slice To value
   of 90

6. Name the tube Grain Chute, and assign
   it any color you want

Radius 1:	3'0.0"
Radius 2:	2'0.0"
Height:	50'0.0"
Height Segments:	1
Cap Segments:	1
Sides:	10
☑ Smooth	
☑ Slice On	
Slice From:	270.0
Slice To:	90.0

Cylinder

7. Make a Cylinder to represent a hole in
   the side of the silo, for the grain chute:

8. Name the cylinder Fake Hole, and
   assign it the black custom color

Radius:	3'0.0"
Height:	1'0.0"
Height Segments:	1
Cap Segments:	1
Sides:	12
☑ Smooth	
☐ Slice On	
Slice From:	0.0
Slice To:	0.0

9. Lastly, make a box for a barn door:

Length:	10'0.0"
Width:	7'0.0"
Height:	0'3.0"
Length Segs:	1
Width Segs:	1
Height Segs:	1

10. Name the box Barn Door, and assign it Custom Color #1, the silver

11. Save the scene

The Asset Browser

The Asset Browser is a floating window that displays thumbnail images of models (in many formats) and bitmaps from various sources, and allows you to drag-and-drop from the browser into your scene. You can pull resources in from your hard drive, from a CD, and even from a collection of Web sites set up to work with the VIZ Asset Browser.

The last couple items needed for the scene are a cow and a tractor. These have been made for you (entirely from Primitives), and you can use the Asset Browser to drop them into the scene.

1. Make the Top view active

2. Choose the Utilities Panel (the hammer)

3. In the Utilities Panel, choose Asset Browser

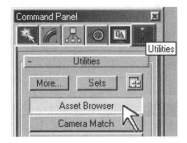

⚠ For some reason, the Asset Browser often appears minimized. It is not actually minimized, it just resizes itself down to a small bar which you'll find at the lower-left of the interface. If this happens, place the cursor on the very corner at the lower-right of the Asset Browser title bar and when the cursor shows the Resize icon, drag the Asset Browser to a larger size.

When you open the Asset Browser you'll see a dialog box with a message about licensing. The Asset Browser browses drives in your computer and Web sites for models, scanned images, and other resources, and resources you find on the Web may have restrictions on how they can be used. Choose OK to close the dialog box.

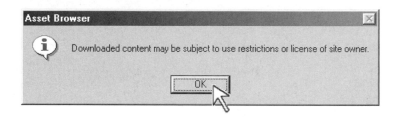

The Asset Browser consists of a browser pane on the left side, a thumbnails pane on the right, and a set of configuration tools above the panes.

The first time you browse to a folder, the Asset Manager will take some time (from a few seconds to several minutes, depending on the number of files in the folder) to create thumbnails of items in the folder. Later, when you browse again to that folder, the Asset Browser will retrieve the thumbnails from a cache on your hard drive, making the process much quicker.

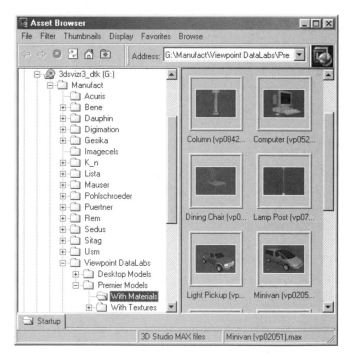

4. Size and position the Asset Browser so you can see the Top view

Drag in a Cow and Tractor

1. Browse to the C:\Viztutorials\Chapter2 folder

2. From the menus at the top of the Asset Browser, choose Filter / 3D Studio MAX files

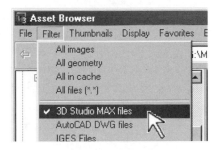

The Asset Browser shows some strange behavior as you drag and drop items. The first oddity is that as you begin to drag a thumbnail from the browser, you will see the international symbol for "No". Ignore this symbol- the .max file will have no problem being dragged into your scene.

The second odd behavior occurs as you drop. With the mouse button still held down, you won't see a ghost image of the .max file at your cursor. When you release the mouse, then you'll see the ghost image. Move the mouse to position the object in your scene, and click to set the object down.

The third quirk is that the Asset Browser windows does not stay floating over the VIZ interface after you drop a resource into the scene. It remains open, but to drop another item into the scene, you will have to select the Asset Browser tablet at the Windows taskbar (or use Alt-Tab) to bring the Asset Browser forward again. Hopefully this behavior will be changed in the next release of VIZ.

3. Drag Cow.max, then Tractor.max into the Top view. Position the cow and tractor near the Grain Pile

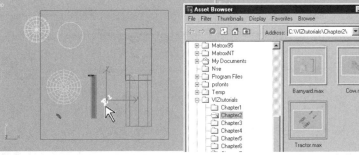

Do not be concerened if you see a message concerning "Obsolete File Format". Just choose OK

4. Close the Asset Browser

5. Save the scene

Transforms

Now you will begin assembling the barnyard parts, first moving and rotating some objects just by eye, and then positioning others using some tools for accuracy. Move and Rotate are two of three tools called Transforms, the third is Scale. To Transform an object is to change its position, orientation, or size, without changing the structure of the object. All the vertices of an object are in the same relation to each other after a Transform as they were before. The object is simply in a different place, a different orientation, or is a different size. Tools that do change the structure of an object are called Modifiers.

Positioning "By Eye"

If you are a user of CAD programs, you have been conditioned to never place things just by eye; things must be placed accurately using snaps or typed coordinates. In 3D Studio, placing things by eye is often acceptable. In this scene, you're creating a visualization, not construction documents, and "if it looks good, it is good". Begin assembling the barnyard by visually positioning the chute and the fake hole relative to the silo:

1. Select the Grain Chute by clicking on it

2. To prevent accidentally deselecting the Chute or selecting something else, hit the spacebar on your keyboard. At the Status Line you will see a padlock turn yellow. You can also click that icon to lock and unlock selections, but the spacebar is faster

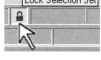

The Chute needs to be rotated and moved in two views: the Left view and the Top view. Which view you work in first matters. If you rotate in the Top view first, then switch to the Left view, you'll be viewing the Chute at an angle, and rotation will not be predictable. Start with the Left view, then switch to the Top view.

3. Make the Left view active, Zoom Extents

4. In the toolbar along the top of the interface, choose the Select and Rotate tool

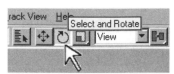

With the Rotate tool active, a red, green, and blue axis icon appears at the left end of the Chute. This is called the Transform Gizmo. It allows you to constrain a Move, Rotate, or Scale action to a particular axis.

5. Position the cursor over the blue axis of the Transform Gizmo, which appears as a blue dot, since you're looking straight down the axis. When you see the Rotate cursor, drag upward to rotate the Chute down as shown here:

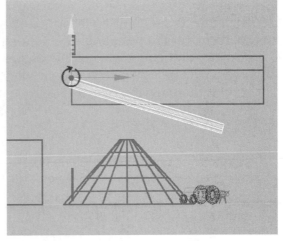

6. Switch to the Top view

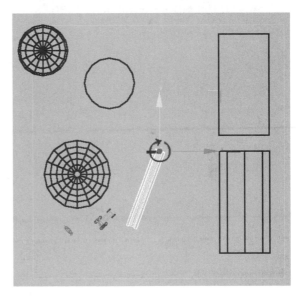

7. In the Top view, Rotate the Chute to match the angle between the Silo and the Grain Pile, as shown here:

8. Switch from the Select and Rotate tool (which from here on will simply be referred to as the Rotate tool) to the Select and Move tool

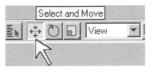

9. With the cursor showing the Move symbol, place the cursor over the right-angle corner of the Transform Gizmo (dragging by the corner allows free movement in the XY plane) and position the Chute so that the upper end is buried in the Silo, as shown in the image below:

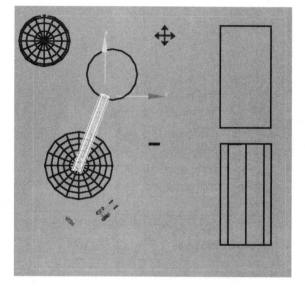

If the lower end of the chute is not directly over the top of the grain pile, that's ok. Move the grain pile after you position the chute

10. Rotate the Chute again if needed

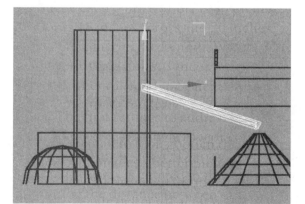

11. Switch to the Left view, position the cursor over the vertical (Y) axis of the Transform Gizmo, and set the height of the Chute so that its low end is a bit higher than the top of the Grain Pile

Using the same techniques as for the Chute, now position the Fake Hole so that it appears to be a hole in the Silo from which the Chute drops.

1. Unlock the Chute selection (hit the spacebar)

2. Select the Fake Hole, then relock selection

3. First In the Top view, and then in the Left view, move the Fake Hole to near where the Grain Chute meets the Silo

4. Get a close-up view of the Fake Hole. Choose Zoom Extents All Selected so the hole fills all the views,

then use Zoom All to zoom out a bit in all views, as shown below:

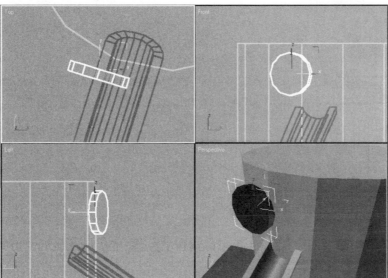

5. Using Move and Rotate in the Top and Left views, position the Fake Hole exactly

6. Unlock the selection

7. Save the scene

Positioning Using Object Snaps

When things need to be positioned accurately, you have three tools available: Object Snaps, the Align Tool, and Transform Type-In. If you are an AutoCAD user, you will find that 3D Studio's Snap tools take some getting used to. By way of introduction, place the Silo Cap on top of the Silo using Snaps.

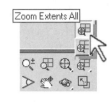

1. Zoom Extents All to see everything in all views

2. Type H, and select the Silo Cap from the Select Objects dialog box

Snap Settings

3. Near the bottom of the interface, at the Status Line, activate the Snap button

4. Right-click over the Snap button to bring up the Grid and Snap Settings dialog box:

5. Uncheck Grid Points, and check Vertex

6. Select the Options tab, and in Options, uncheck Use Axis Constraints. If this box is left checked, you won't be able to lift the Silo Cap off the ground plane

7. Close the Grid and Snap Settings dialog box

When your cursor is over geometry in the active view, you should now see a cyan cross at vertices of objects

Working in four small views, it is difficult to select and snap on to the correct vertex; work in one large Perspective view.

1. Make the Perspective view active, and type the letter W.

2. If the Perspective view is in shaded mode, right-click over the word Perspective found in the upper-left of the view. From the resulting pop-up menu, choose Wireframe. Do you remember the keyboard shortcut for this? It is the F3 key.

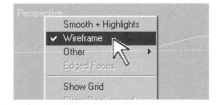

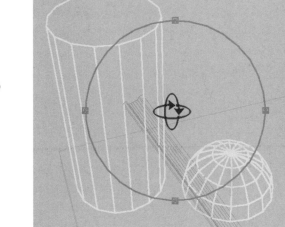

3. Zoom, Pan, and Arc-Rotate to get a good view of the Silo and the Silo Cap

4. Choose the Move tool

5. The Silo Cap should still be selected. If it isn't, select it

6. Move the cursor over the Silo Cap, and look for a cyan cross on a vertex along the bottom of the hemisphere, somewhere on the foremost edges

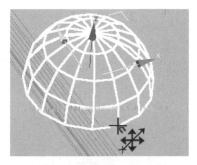

7. With the Snap marker showing, click-hold, and drag the Silo Cap up to the top of the Silo. As your cursor nears the vertices on the top of the Silo, the Snaps should take effect, and the Silo Cap should snap onto the Silo. Match the vertices so that the Silo Cap sits perfectly atop the Silo

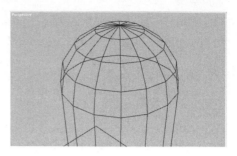

8. Turn off the Snap button

Using snaps in a perspective view is a bit odd, as the object being dragged jumps around and changes size as the cursor passes over eligible snap points that are closer to or further from the viewpoint. The process is more orderly in views without perspective, so you probably should change perspective views to User views before using snaps.

 There is a limitation to the use of snaps in VIZ3, and it involves Backfaces. VIZ objects consist of triangular faces. Each face has a side that accepts materials and shows up in the viewports and the rendering, and a side that does not. The side that renders is called the Normal face, and the invisible side is called the Backface. Look at the bottom of the Silo in the Perspective view; you see only the front edges. Disregarding backfaces makes the program faster, both in rendering and in viewport display. The display can be set so that you see both normals and backfaces, and the renderer can also be set to render both sides. But, no matter how you configure the views and renderer, *y o u cannot snap to backfaces*. This means that any faces that point away from your point of view are ineligible to be snapped to. This fact sometimes makes VIZ snaps somewhat difficult to use.

When you're done using the Snap tool, turn it off. The snap tool interferes with the function of some other tools, most notably the Transform Gizmo.

Positioning with the Align Tool

In many situations the Align tool is a faster, easier way to assemble things accurately. Use the Align tool to set the Barn Roof on the Barn. You will first need to get a good view of the Barn using Arc Rotate:

1. Type H and from the Select Objects dialog box, select the Barn and Barn Roof. To select multiple items from the list, hold the Ctrl key as you select

2. Zoom Extents Selected

3. Arc Rotate Selected in the Perspective view to set a point of view from above, that gives a good view of the Barn and Barn Roof. Remember to start the Arc-Rotate dragging inside the green circle- starting the drag motion outside the circle rolls (tilts) your point of view, which you probably don't want to do

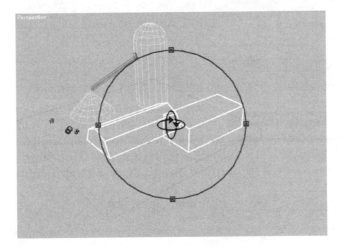

4. Type H, and select just the Barn Roof

5. At the left of the interface, from the vertical toolbar, select the Align tool

6. With the cursor showing the Align icon, click on the Barn. The Align Selection [Barn] dialog box appears. Position the dialog box so you can see the Barn and Barn Roof

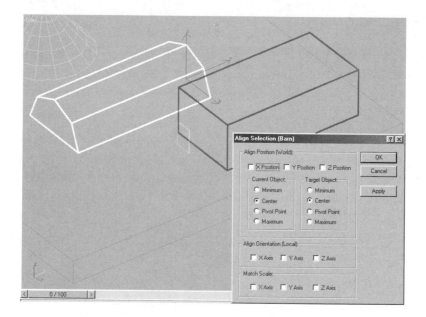

7. In the Align dialog box, put checks in the check boxes for X Position and Y Position. The Barn Roof will align in two directions, so that it is directly above the Barn. Note the settings in the Current Object and Target Object groups-both are set to Center

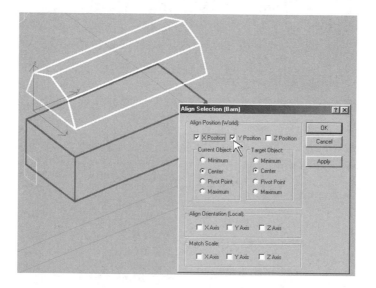

The Align tool uses objects' local axes to align things. There are two axis systems at work in a 3D Studio scene; a set of axes that order the general environment, and a set of local axes assigned to each object. Local axes get assigned and oriented as objects are made (this is also true in AutoCAD, but in AutoCAD you cannot see or manipulate the local axis). Getting quick with the Align tool requires that you learn to view the Axis Icon for an object and understand what point on the object represents the Minimum X, the Maximum X, the Minimum and Maximum Y, and the Minimum and Maximum Z.

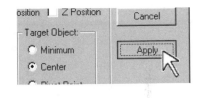

8. At the right of the dialog box, select the Apply button. The X Position and Y Position check boxes will clear

9. Check the Z Position check box. The bottom of the Barn Roof is the roof's Minimum Z. The top of the Barn is the Barn's Maximum Z. So set the Current Object choice to Minimum, and the Target Object choice to Maximum. The Barn Roof should now sit perfectly on the Barn

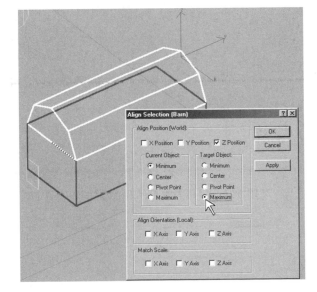

10. Choose OK to close the Align dialog box

11. Type H, and select the Barn Door

12. Use the Align tool to set the Barn Door on the front of the Barn, with the right side of the door (Maximum X) centered on the barn. Make sure you understand why the front of the door represents the Minimum Y, and the back side of the door represents the Maximum Y. You are trying to place the back of the door against the front of the barn, so the choices for the Y Position are Current Object Maximum, Target Object Minimum. Don't forget to address Z Position

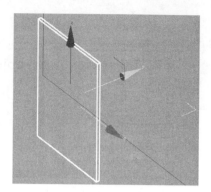

13. Once you've Aligned the Barn Door to the Barn and closed the Align dialog box, type the F3 function key to set the Perspective view to shaded display mode

14. Zoom Extents to see the entire scene

At this point, your scene should look something like this:

If it doesn't because you created objects in different locations, use the Move and Rotate Transforms to rearrange the scene. Work primarily in the Top view. Make sure the Snap tool is off before moving anything.

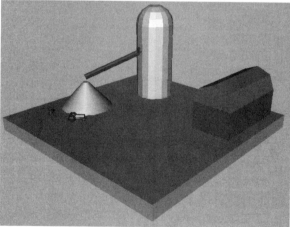

Before moving or rotating the Silo assembly, make it into a Group, as described in the following section.

Making a Group

Because the Silo, Silo Cap, Fake Hole, and Grain Chute are in the correct relation to each other, you would not want to ruin that arrangement if you need to move the entire silo assembly. Group the parts:

1. Type H, and from the list select Silo, Silo Cap, Fake Hole, and Grain Chute. To make multiple selections from the Select Objects dialog box, hold the Ctrl key on the keyboard down as you select

2. With the parts of the assembly selected, choose, from the menus, Draw
/ Group / Create

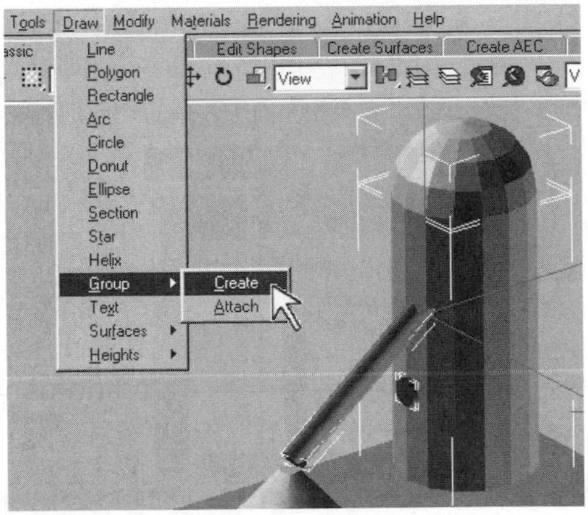

3. Name the Group Silo Assy, and choose
OK

If you later need access to individual parts of that Group, you can select it and choose Modify / Group / Ungroup, or Modify / Group / Open (to temporarily ungroup it).

Now that it is grouped, the Silo Assembly will act as one object.

Duplicating Objects

The cow looks lonesome, and we need a second barn door and another tractor. Duplicating is very simple in VIZ- you move an object while holding down the Shift key on the keyboard. Use this method to make more cows:

1. Type W to return to four views

2. Make the Top view active, then type W to maximize the Top view

3. Type the letter H, and from the list, select Cow01

4. Select the Move tool

5. Make sure Snap is off

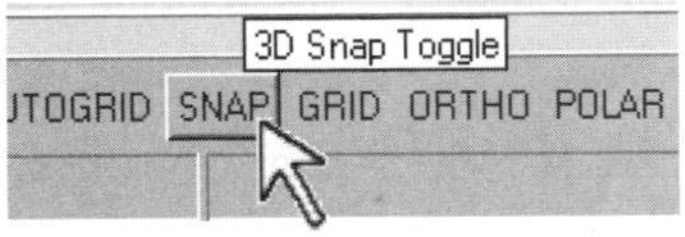

6. In the Top view, position the cursor over the corner of the Transform Gizmo, hold the Shift key on the keyboard, drag a short distance toward the upper-right, and release. The distance that you drag will be used as the spacing between cows in a multiple array, so don't drag too far.

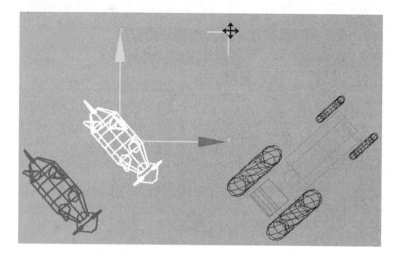

As you release the mouse button, the Clone Options dialog box appears:

7. Set the Number of copies spinner to whatever number of additional cows you want in your scene. Note that the new cows will be named using a number sequence: Cow02, Cow03, etc.

8. Set the Object group to Reference, choose OK

The Clone Options dialog box introduces a key concept in VIZ: Copy versus Instance versus Reference.

A **Copy** is a duplicate that has no ties to the original, and no hierarchical relationship. Making numerous copies of an object can greatly increase file size.

An **Instance** is exactly like an AutoCAD block; it is another occurrence of the object, and if you modify one instance, you modify them all. There is no heirarchical relation among instances. Using instances instead of copies will keep your file size down.

A **Reference** does have a hierarchical relationship. The original object is the Parent, and references of that object are the Children. Any modifier applied to the parent will also modify the children. But a modifier applied to a child will not affect the parent, nor will it affect the other children. Anytime you see groups of similar (but not identical) objects in a VIZ scene, chances are good that those are References.

In the case of these cows, suppose you want to make some calves without horns, and suppose you want to give each cow a unique twist to its tail. That eliminates Instance as a possible object type choice, since removing horns and twisting the tail of any one of the set of Instances would do so to all of them identically. You could use Copy, to have the freedom to modify any cow any way you wanted, but then suppose that you wanted to add eyes and mouth to all the cows; you would have to add them repeatedly for every copied cow in the scene. The best choice here is Reference, so that adding eyes and mouth to the parent cow will also add them to all the children cows, and you still are free to modify the various children cows uniquely.

The word "modify" is used deliberately in this explanation; Modifiers are different from Transforms (Move, Rotate, Scale), and Transforms work differently on Instances and References. Transforms ignore the Instance or Reference relationship. Scaling one cow in a set of Instanced cows won't scale any others.

You should now have a herd of cows:

9. Use the same Shift-Move technique to make a second tractor

Should that second tractor be a Copy, an Instance, or a Reference? If you were to modify the design (the structure) of either tractor, you would probably want the modification to affect the other one identically, so the best choice would be Instance.

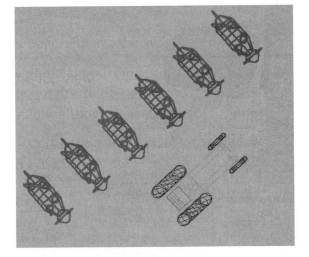

10. Use the Transforms (Move, Rotate, Scale) to arrange the cows and tractors in the scene, and to scale some of the cows larger and smaller. After scaling a cow, you will find that its hooves are either off the ground, or buried in the ground, so you will need to move the scaled cows in the Front or Left view

11. Type W to return to four views

12. Make the Front view active

13. From the View Navigation tools choose Region Zoom, and use it to zoom in on the front of the barn. To use Region Zoom, click-hold at one corner of the area you want to magnify, drag to the diagonal corner of the area to be magnified, and release

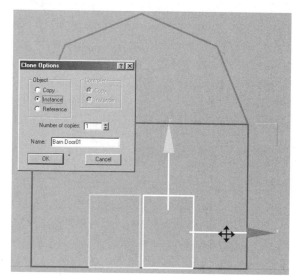

14. Type W to maximize the Front view

15. Using the Shift-Move technique, make a second Barn Door to the right of the first one. Set a gap between the two doors. In the Clone Options dialog box, choose Instance, set Number of copies to 1, then choose OK

16. Type W to return to four views

17. Zoom Extents All

18. Save the scene

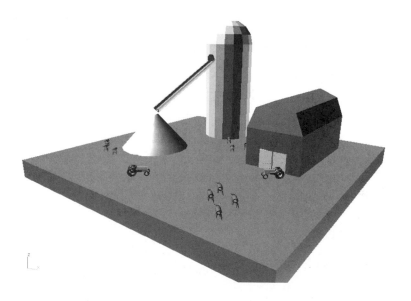

The barnyard needs a couple more geometric objects: some trees, and a fence around the pile of grain. Once those are in place, you'll finish the model by adding a background and some lights to cast shadows.

AEC Objects: Railing

VIZ 3 features several parametric objects for architects: Wall, Door, Window, Stair, Railing, Foliage, and Terrain. Most of these parametric objects are fairly simply configured (the doors and windows, for example, don't have hardware), and in a scene where more detail is required, you will likely want to add geometry to customize them. But "out-of-the-box" they are great templates and great time-savers.

Build a fence around the pile of grain.

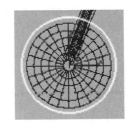

1. Make the Top view active

2. In the Create panel, choose the Shapes button, then choose Circle

3. In the Top view, click-hold at the center of the Grain Pile, and drag out a circle slightly larger than it

4. In the Create panel, choose the Geometry button, open the subcategory drop-down list and choose AEC Extended

5. Under Object Type, choose Railing

6. Under the Railing parameters, choose Pick Railing Path. Position the cursor over the Circle, make sure you see a vector symbol at the cursor, and click. The result won't look like much yet, but you should see a couple dots on the circle (they are fence posts)

7. Leave the Segments spinner at 1

8. Set the Top Rail Profile drop-down list to Round

9. Set Top Rail Depth to 6"
   Top Rail Width to 4"
   Top Rail Height to 5'

10. Leave Lower Rail Profile square
    Set the Depth to 4"
    Set the Width to 3"

11. Check Generate Mapping Coords

12. Open the Posts rollout
    Leave the Profile square
    Set Depth to 6"
    Set width to 6"
    Set extension to 1'

13. Open the Fencing rollout, set Type to None, then close the Fencing rollout

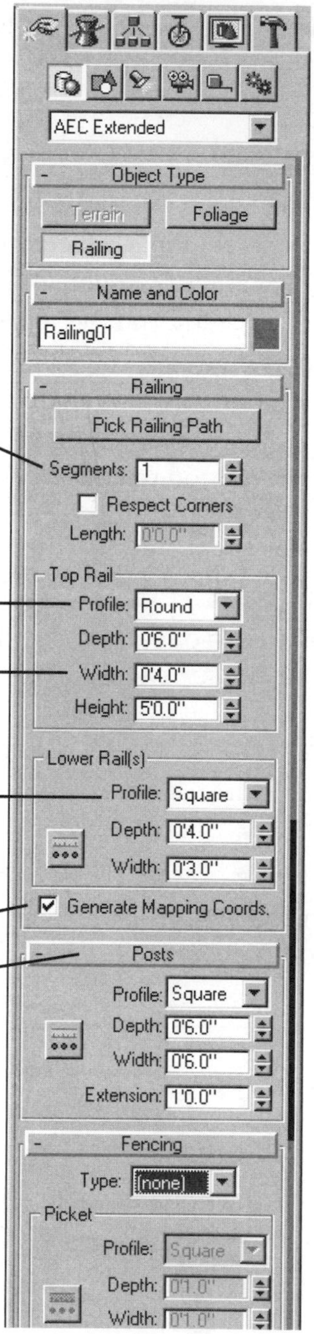

14. In the Lower Rail(s) group, choose the Lower Rail Spacing tool

15. Set the Count spinner to 3. Open the drop-down list for spacing and division, and view the numerous options. You'll need this much flexibility when building the railings for an interior staircase. Leave the drop-down set to Divide Evenly, No Objects at Ends, and close the dialog box.

16. Under the Posts rollout, choose the Post Spacing tool

17. In the Post Spacing dialog box, set the Count spinner to 12, leave everything else as is, and choose Close

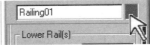

18. Assign a brown color to the Railing object

19. Zoom Extents All Selected to zoom tight to the Railing in all views

20. Make the Left view active, and choose Arc Rotate Selected

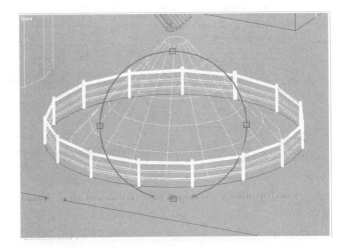

21. Arc Rotate to get a good look at the Railing

The path for this Railing object was a 2D circle, but you can just as well use a path that has a change in height along it, such as a line around the rolling hills of a horse farm, or a line that follows the carriage of a stair. VIZ's parametric Stair objects, which you will work with in the next chapter, have these railing paths built into them, which makes creating staircases and laying out their posts, railings, and ballusters a fairly quick task.

Railing Terminology

Rail- The horizontal elements

Top Rail- The larger, uppermost horizontal element

Lower Rails- One or more narrower horizontal elements

Posts- The larger vertical elements

Fencing- The narrower vertical elements (pickets) or solid panels, like Plexiglas

Don't let the name "Railing" limit how you use this tool. When you think about it, many items in the built environment are basically railings. The objects below are modeled entirely from Railing objects.

AEC Objects: Foliage

There are a few approaches to populating a scene with trees and shrubs. Each method has limitations.

Many visualizations feature "cutout" or "billboard" trees and shrubs. In this method, the geometry is nothing but a rectangular face. A picture is applied to that face in two ways: once to display a photo of a tree (diffuse map), and once to determine where the rectangular face should be seen, and where it should be invisible (opacity map).

Diffuse Map

Opacity Map:
White = Opaque,
Black = Transparent

The cutout technique keeps model sizes small and renders quickly, but it limits the subtlety you can get in shadows, and it doesn't work well in animations, where you expect truly 3D foliage.

When 3D Foliage is needed, there are three approaches: fully modeled trees and shrubs, procedural (mathematical) foliage, and 360-degree photographs (the Realtrees plug-in).

With procedural foliage, the geometry in your scene is limited to the trunk and branches, and at rendering time leaves, blossoms, pinecones and such are added by the renderer, usually using random mathematics. This method can make great-looking foliage, but uses huge amounts of computing power.

The Realtrees plug-in places one of 360 photos of a tree into your scene, at every frame of animation, depending on the camera's point of view.

VIZ 3 features fully modeled, parametric foliage. Leaves are mapped onto thousands of polygons, using the opacity mapping "cutout" technique described previously. For display speed, the foliage can be viewed in several modes, and the display of leaves and branches can be turned off.

Creating Trees

1. Make the Top view active and Zoom Extents All

2. In the Create panel, under the Geometry category (button) and the AEC Extended subcategory (drop-down list), choose Foliage

3. Scroll through the thumbnails, find a tree you like (elm, oak, and weeping willow are good choices for this scene), and drag it over the Top view. Don't release the mouse button yet

4. As you move the mouse to position the tree before releasing the mouse button, move slowly. Geometric trees can overwhelm even a fast computer. Release the tree anywhere in the southeast corner of the barnyard

5. When you've dragged in the first tree, open the Parameters rollout at the bottom of the Create panel

6. In the Parameters rollout, set the tree Height to 40′

7. Set the Density to .5 and set the Pruning to .2 (pruning removes the lowest branches)

8. Next to the Seed spinner, choose the New button a few times. The tree takes on a different, random shape with each new Seed number. Many VIZ objects and modifiers feature Seed numbers – random mathematical generators

9. In the Show group, uncheck Leaves and Branches to see the effect, then check them again

10. In the Viewport Canopy Mode group, choose Always. You will see the leaves represented as a translucent fabric in shaded views, and as a web in wireframe views. Set the Viewport Canopy Mode back to When Not Selected

11. In the Level-of-Detail group, alternate between High, Medium, and Low, just to see the result, then leave it set to High

12. Drag a couple more trees into the Top view, arranging them so your scene looks something like the following image:

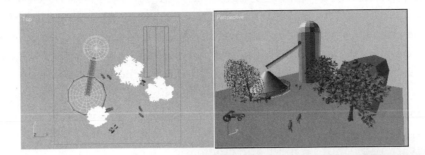

13. Make the Perspective view active, then use Arc-Rotate, Zoom, and Pan to get a viewpoint similar to this one:

14. Choose the middle Render teapot, Quick-Render

15. Save the scene

As you can tell from this exercise, populating a scene with 3D trees is not practical if the scene contains more than a dozen or so trees. Usually in still-image renderings, a few cutout trees that are critical to the scene are placed in the VIZ scene, and the rest of the background foliage is added later in a paint program after the VIZ image is rendered. For animations, one of the best solutions for foliage to emerge recently is the "360 cutout" method, such as the package offered by Archvision called Realtrees (www.archvision.com). In this method, a subject (Archvision offers foliage and people) is photographed from 360 angles, and a stand-in object is placed in the VIZ scene. Depending on the VIZ camera's point of view, the Realtrees plug-in program retrieves the photo that corresponds to the viewing angle. As the camera moves through the scene, a new photo is retrieved at every frame of animation, resulting in a 3D photograph of a tree, shrub, or person.

Samples of Realtrees foliage are included on the Designer's Toolkit CD that ships with 3D Studio VIZ. You can drag Realtrees and Realpeople into your scenes from the Asset Browser.

Creating Cameras: The Target Camera

To establish a permanent perspective view that will not be affected by view navigation tools, create a Camera. Like Cylinder and Box, a Camera is a parametric object; its parameters are always available for editing in the Modify panel. A VIZ Camera shares some characteristics with a movie camera: zoom lens length, field of view, dolly and truck tools, even a tool for mimicking one of Alfred Hitchcock's favorite camera tricks. VIZ Cameras do not feature an aperture setting or shutter speed- the "exposure" of a rendering is determined entirely by the qualities of the lights and materials in the scene, not by Camera properties. There is a set of tools found in the Rendering / Effects menu that allow a variety of photographic and cinematic effects to be added to the image in a second pass after the scene is rendered initially. If you want to add depth of field, blurring from movement, lens flare, or film grain to your image, VIZ 3 has tools for doing this. For now, just drag out a Camera object, choose Camera settings, and compose a good view of the barnyard.

1. Close the rendered view

2. Make the Top view active, and Zoom Extents

3. In the Create panel, choose the Cameras category, then choose the button labeled Target

4. In the Target Parameters rollout, from among the Stock Lenses buttons, choose the 20 mm lens

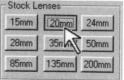

5. In the Top view, drag out a Camera. Your first click-hold places the Camera, then drag, then release to set the position of the Target. Set the Camera's position near the midpoint of the bottom edge of the Barnyard Base, and drag the Target to the center of the barnyard:

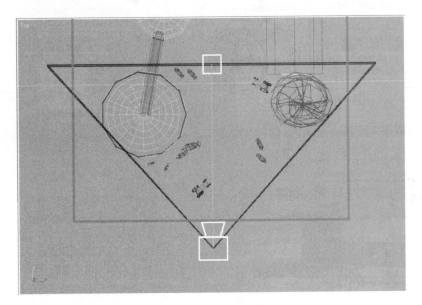

6. Make the User view (the lower-left viewport) active

7. Type L to restore that view to a Left view, then Zoom Extents the Left view

8. Make the Perspective view active, then type the letter C to look through the Camera

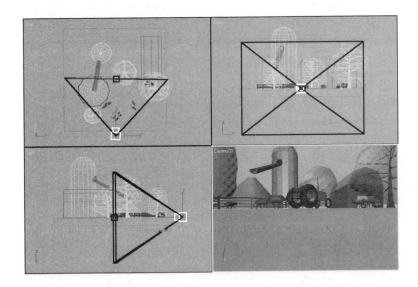

Selection Filter - Cameras

You will be moving the Camera icon in various viewports to adjust the view. With this much geometry in the scene, it is very easy to accidentally select things besides the Camera as you work. To prevent this:

1. In the toolbar above the views, open the Selection Filter drop-down list, found just to the left of the Select by Name button

2. Set the Selection Filter to Cameras. You can now select only the Camera and its Target

Composing the Camera View

There are two methods for composing the view through the camera; either use the Move tool in the various views to move the camera icon and the target icon, while watching the result in the Camera view, or make the Camera view active, and use the view navigation tools specific to Camera views to adjust the composition.

If you've never studied drawing, painting, or photography, you ought to get a book on the basics of good scene composition. A few basic ideas likely found in such a book are:

Divide the scene in thirds. If there is a strong horizon visible, set the horizon roughly a third of the way or two-thirds of the way up the view. Keep the two-thirds guideline in mind for the horizontal composition as well.

Do not center the primary subject in the view- it is more interesting if the primary subject is a bit off-center.

In exterior eye-level views, lift the target a bit higher than the camera. We look slightly up when we survey our surroundings, and the slight perspective distortion caused by looking up makes a more dynamic composition.

The composition needs objects in the foreground, middle ground, and background. Let closer objects partially obscure further objects. Let objects touch the edges of the picture, and run off the edges.

Try to suggest a line of action at an angle to the picture plane

Composition concepts applied:

Moving the Camera Icon

1. Make the Left view active

2. Select Camera01.Target

3. Move the target up slightly, just a few degrees, then select both the camera and the target and move them both up roughly ten feet

4. Switch to the Top view, and Move the Camera left a bit, as shown in this image:

Your composition probably does not look just like the Camera view shown on the preceding page. In that image, objects in the scene were moved about a bit to improve the composition. Feel free to do the same, working in the Top view and watching the result in the camera view.

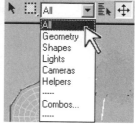

5. Set the Selection Filter drop-down list back to All

6. To improve the composition, move objects in the Top view while watching the composition through the camera view

7. When the composition is to your liking, save the scene

Multiple Camera Views

If you need to set up a dozen different views of a scene, it is probably a better idea to animate a single camera's position over twelve frames of animation than to create twelve Camera objects in your scene. Not only does one animated camera reduce clutter in the viewports, it also makes rendering all your views easy; just set the renderer to render the first twelve frames of animation as individual still images. You can set this up at the end of the day, and you will have your twelve views in the morning, with no further prompting required. Establish a second view of the barnyard using this method.

1. Make the Camera View active

2. Turn on the Animate button

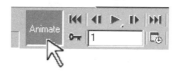

3. Below the views, click once on the arrow at the right of the Time Slider to advance to Frame 1 of the animation

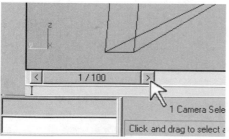

Camera Navigation Tools

Notice that the view navigation tools have changed to a set pertaining to Cameras:

Dolly Camera: In cinematography, the dolly is the cart that runs on small train tracks. To Dolly Camera is to move the Camera straight toward or away from the Target. The Target does not move.

Dolly Target: The opposite of Dolly Camera- the Camera stays still, and the Target is moved straight toward or away from the Camera.

Dolly Camera + Target: Both move along the straight line that runs from Camera to Target

Perspective: Dollies the Camera, while simultaneously changing the Field of View, resulting in a distortion of perspective in an unchanging composition. Most Alfred Hitchcock films use this effect at some point.

Roll Camera: Tilts the Camera

Field-of-View: Adjusts the Camera's zoom lens. A short lens (15 mm) gives a wide FOV, and greater perspective distortion. A long lens (300 mm) gives a narrow FOV, and less perspective distortion. It is better to set lens length in the Modify panel than to rely on this button.

Truck Camera: Moves the Camera and Target sideways. In CAD programs, the word "pan" is usually used to describe this action

Orbit Camera: The Target stays still while the Camera circles around it.

Pan Camera: A true pan, in cinema, is when the Camera stays still and the Target moves in an arc about the Camera, sweeping across the scene.

Compose a View Through-the-Lens

4. With the Animate button on, the Time Slider at frame 1, and the Camera view active, use the Orbit Camera tool to circle around to the right some:

5. Use Dolly Camera to move the Camera toward the Target, until the Barn is almost out of the picture:

6. Use Truck Camera to move the Camera and Target toward the left. Use whatever combination of Camera navigation tools you need to compose a good view that's similar to the one shown here. Move objects in the scene a bit to improve the composition if you want.

7. When the view is set, turn off the Animate button and save the scene

8. Alternate the Time Slider between frames 1 and 2 to review the two compositions, then leave the Time Slider at Frame 0

9. Render the Camera view

Background Image

The scene is a bit gloomy. A background other than black would certainly help.

1. From the menus, choose Rendering / Environment

2. In the Environment dialog box, choose the Environment Map bar, labeled None

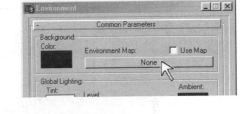

3. In the Material / Map Browser, select Bitmap, then choose OK

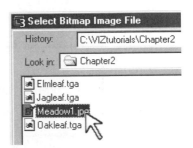

4. In the Select Bitmap Image dialog box, navigate to the C:\Viztutorials\Chapter2 folder. Select Meadow1.jpg for the background image, then choose Open

5. Close the Environment dialog box, save the scene

6. Quick-Render the Camera view

Definitely an improvement with the background image. There are two things conspicuously missing from this image. One is realistic materials, which you will begin to use in the next chapter, and the other is shadows, which you will add next..

Exterior Lighting: Basic Scheme

The last thing you will do to this scene is add a couple lights to get shadows and a more dramatic illumination.

If you are rendering exterior still images, and your views are limited to three sides of the model, you can probably light the scene correctly with four lights, and certainly with five. You need a light to act as the Sun and cast shadows, you need two fill lights so that surfaces in shadow are not totally dark, and you need a light from below to illuminate soffits and such. Beyond that, any additional lights you add to the general illumination scheme will probably do more harm than good.

VIZ lights do not attempt to correlate to the real photometric properties of natural or man-made light. Properly lighting a VIZ scene is more art than science, and is one area in which the clever tricks picked up over the years separate the competent users of VIZ from the masters. It is vital to keep in mind, as you develop your lighting scheme, that the properties of the Light objects are only a part of the solution; the properties of the materials you develop are just as much a part of the lighting solution. Don't spend hours obsessing over light parameters without opening the Material Editor.

A lighting scheme for the exterior of a building should work something like this:

If the rendered buildings are to be photocomposited into a site photo, then the shadow-caster needs to be positioned so that it casts shadows that will agree with shadows in the site photo. Otherwise, picture the rendering in your mind and position the shadow-caster to create interesting, dramatic shadows that will improve the viewer's understanding of the forms in the scene. The shadow-caster is usually the brightest light in the scene, and should give off slightly yellow light.

The first fill light illuminates surfaces that face away from the shadow-caster. It is placed slightly above the ground. The second fill light is positioned to light the same surfaces as the shadow-caster, and it is also slightly above the ground.

Light Below
Value = 80

Fill Light
Near Ground
Value = 180

Cameras

Fill Light
Near Ground
Value = 225

Shadow-Caster
High Angle
Value = 255

CHAPTER 1

The VIZ Process

This chapter is an overview of the process of building a scene: modeling objects, positioning and duplicating objects, creating cameras, adding lighting, assigning materials, and rendering.

﹡ Modeling the Wine Bottle, Glasses, Candlestick, and Table Base: these are all modeled by drawing a 2D profile shape and applying a Lathe modifier to the shape to spin the shape about an axis and create a 3D object. See pages 187–197 for practice drawing shapes and lathing them.

﹡ Modeling the Fruit Dish: two shapes combine to make the dish: a star shape lying on the table, and a profile shape along which the star is extruded using a Bevel Profile modifier (pages 178–186).

﹡ Materials: except for the fruit, these are materials from pre-made VIZ libraries. See pages 32–34 and pages 129–132 for practice using material libraries.

CHAPTER 2

The Barnyard

This chapter teaches the basics of building a scene from the ground up: object creation, efficient use of the VIZ interface, accurate positioning and assembling of objects, a scheme for exterior lighting, and the establishing of a good composition through the camera.

﹡ Trees: there are several options for adding trees. This scene uses the 3D trees available in VIZ. Options for foliage are described on pages 67–69 and 287–288.

﹡ Background Image: the steps for adding a background image are on page 77. In this scene, shadows are present against the background image. This is done by assigning a Matte / Shadow material to the base of the scene, which makes the base transparent, but still able to receive shadows. See pages 82–84 for practice with the Matte / Shadow material.

The Tower and Furnishings

Chapter 3 introduces the parametric AEC objects: Wall, Door, Window, Stair, Railing. These objects are built by specifying their parameters. For example, for stairs you specify the type of stair (straight, L-shaped, U-shaped, spiral), the tread dimensions, riser height, handrail height, stringer and carriage dimensions, and so on. As the design evolves, you simply edit the parameters of the AEC objects to update the components of the building.

The furnishings shown in this image are built in Chapter 4. The tools used in making furniture are the same powerful modeling tools you will use to make a wide variety of objects that might be produced by any of the design professions. The leather seats are an exercise in mesh editing (pages 158–169). The table is made with the Bevel Profile modifier (pages 178–186). The cafe chairs are made mostly from Loft objects (pages 198–213).

CHAPTER 5

Landscapes and Natural Materials

Modeling landscapes is often more challenging than modeling buildings. The materials needed in a landscape will be some of the most complex materials you will design in VIZ.

* The Pond: without it this image would be fairly dull. The basic appearance of a water surface is actually rather easy, but the transition from water to sandy shore to grass is a bit tricky. See pages 271–286 for practice making this pond.

* The Grass: the key to nice grass is to start with a couple of good scanned images of grass (golf courses are the best source of grass maps) and to then blend the two images in a random pattern so that no seams at the edges of the photos are allowed to appear. See pages 250–268 for the steps to make this grass.

* The Trees: trees are placed using the Scatter tool. See pages 287–294.

CHAPTER 6

Lighting the Tavern

The materials in the scene for this chapter are left extremely simple (just colors) to point out that good light and shadow do more for a scene than anything else.

✳ **Globe Lights:** the two lights over the bar make use of Attenuation—the diminishing of light intensity with distance. Nearly every light in an interior lighting scheme needs attenuation. See pages 305–306.

✳ **Neon Sign:** the tubes of the sign are modeled using Renderable Splines (page 339). The glow around the tubes is an effect added in a second rendering pass. See pages 337–343 for practice with the Glow effect.

✳ **Shadows on Floor:** the shadows from the barstools are being cast by a light that casts no illumination. See pages 351–353 for practice with shadow–only lights.

CHAPTER 7

Man-Made Materials

The Material Editor is the most complicated part of VIZ, and also the area calling for the most creativity. Much of the detail that you see in a rich, complex scene is not built into the geometry, but rather is painted onto very simple surfaces using simple materials techniques.

✳ **The Floor:** the grout lines are done using Masking. The floor is brown stone except where a bitmap masks the stone texture, revealing the color of the grout. See pages 394–396 for practice with masking. The reflection is done with the Flat Mirror map (pages 399–403).

✳ **The Perforated-Metal Reception Desk:** "never model what you can fake with a map" (Ted Boardman). Modeling the star-shaped holes would be difficult. Instead, an Opacity Map consisting of black stars on a white field is used to make transparent star-shaped areas on a continuous surface. See pages 388–393 for practice making such a material.

The Gallery

✴ Tile Floor: the tile surface uses the Cellular procedural map, rather than a bitmap. The grout lines are done with a bitmap used in two channels—the Bump channel and the Specular Level channel. See pages 394–398 for practice making tile floors.

✴ Carpet: a hybrid procedural map / bitmap technique. The material uses a bitmap, but the bitmap is not a scan of a carpet sample. The bitmap was created by making a four-foot by four-foot plane and applying a material to it that blends and mixes several procedural maps (Cellular, Speckle, Noise) to simulate the carpet.

This procedural material does not blur well into the distance. So the four-foot square was rendered from above to create a bitmap. A new carpet material was made using the bitmap, and the resulting carpet blurs nicely at the far end of the hall. See pages 371–377 for practice making carpet.

✴ Books: these are hundreds of simple 3D books, distributed among the shelves using the Scatter object (see pages 289–293). The book covers are bitmaps scanned from real books.

✴ Pictures on Wall: see pages 193–196 for practice making framed pictures.

Burt Hill Kosar Rittelmann Associates, Boston, MA

4

Richard Saivetz Assoc. / Garo Dev. Corp, Needham, MA

Condominiums

✳ Roof: the material is simple; a bitmap of cedar shakes is used in the Diffuse Color channel. The tricky part is the mapping coordinates. A Mesh Select modifier (page 239) is applied to the roof object, and a selection set made of all roof planes oriented north-south. A UVW Map modifier is applied to that selection of faces. Then another Mesh Select modifier is applied, and all roof planes oriented east-west are selected. A second UVW Map modifier (rotated 90 degrees to the first) is applied to correctly map the east-west planes. See pages 418–420 for practice with the precise placement of UVW Map coordinates.

✳ Siding: if the view will not allow the close scrutiny of the siding, the easiest way to make lapped siding is to repeat a Gradient procedural map up the walls. In the siding material, Gradient is present in the Bump channel and in the Specular Level channel. The gradient is white at the bottom, black at the top, and repeats twenty times up the walls. Where the gradient is white the siding appears to protrude and shine the brightest, and where the gradient is black the siding appears recessed and darker. The result is very convincing lapped siding.

✳ Fascia and Soffit: the straight segments of the fascia are simple extrusions of lines. The angled sections are lofted. See pages 198–217 for practice lofting. The soffits are just lines drawn in the appropriate elevation and extruded.

The Research Room

✳ **Wood:** no bitmaps are used in the wood materials in this room; the wood is done entirely with the procedural Wood map. To understand the settings for the Wood map, picture a section and a plank sawn from a felled tree. The settings pertain to using the map in three dimensions, with Radial Noise controlling distortion across the rings, and Axial Noise controlling distortion along the length of a board. In most architectural applications you are depicting boards, not 3D blocks of wood, so you can concentrate on Axial Noise. To get the right Wood settings, start with the Grain Thickness value, which controls the width of the bands of color. The board shown in this image measures 2.5 feet by 12 feet and has a Grain Thickness of three. Next find the proper rotation of the grain. One of the three Angle spinners will likely need to be set to 90 degrees to get the grain to run down the board—find the correct spinner through trial and error. Next, set the Tiling and the Axial Noise. These two settings work together. Noise sends waves through the wood. The Noise spinners determine the intensity (amplitude) of the waves, and the Tiling spinners determine the wavelength. The higher the Tiling setting, the shorter the wavelength, and therefore the choppier the distortion. Only one of the Tiling spinners will have the desired effect on a board—find the right one through trial and error.

Burt Hill Kosar Rittelmann Associates, Boston, MA

Pharmaceuticals Manufacturing Facility

* Diffusers: the glow around the diffusers in the ceiling is done in a second rendering pass that applies a Glow effect. See pages 337–343 for practice with the Glow effect.

* Reflections: the reflection on the floor is done with the Flat Mirror map in the Reflection channel of the material (see pages 399–403). Since Flat Mirror works only on a single flat plane, the floor slab is divided by Material ID numbers to isolate the top of the slab as a seperate material. See pages 166–167 and pages 239–243 for two methods for assigning Material ID numbers. The reflections on the equipment are done with the

Raytrace map in the Reflection channel. See pages 408–415 for practice with the Raytrace map.

* Railings: created with the Railing object. See pages 105–107 for practice with Railings.

* Safety Tape: the strips of tape are modeled as separate rectangles with a UVW Map modifier applied (see page 129). The stripes are accomplished with a Checker procedural map in the Diffuse Color channel. See page 427 for steps for making the Checker map appear as stripes.

*Symmes Maini
& McKee Associates,
Inc. Cambridge, MA*

School Addition

* Stone Walls: these are modeled with the Bevel Profile modifier (pages 178–186) The walls are assigned two sets of Material ID numbers (pages 166–167); one for the lower parts and a second for the caps of the walls. The caps also carry unique mapping coordinates. A Mesh Select modifier (page 239) is used to build a selection of the caps, and a UVW Map modifier (pages 390–393 and 418–420) is applied to that selection.

* Plaza areas—Red Concrete, Grass, Pavers: the various surfaces of the plaza are defined within the perimeter of the plaza using Shapemerge

(pages 238–245). The plaza is therefore a single object, carrying several Material ID numbers for the various surfaces.

* Grass: two bitmaps of real grass are blended together with a Noise mixer map. This grass is nearly identical to the grass developed on pages 250–266.

* Foliage and People: the trees are done with cutouts placed in the VIZ scene (pages 287–288). The shrubs, grasses, and people (and the shadows from them) were added in a paint program.

Davis Partnership P.C. Architects, Denver, CO

The shadow-caster is usually high in the sky. Its rays have a fairly steep angle of incidence to the roof planes, and illuminate the roof well. But its rays have a shallow angle to the wall surfaces, and cannot illuminate walls well. The second fill light, placed near the ground, has a much more perpendicular angle of incidence to those wall surfaces.

The light below is the dimmest light of the three. Though it usually does not cast shadows, you might allow it to cast very subtle ones, to make rich, complex secondary shadows play in the beams of an exposed soffit, for example. If the ground near the building is of a strong color (grass being the most obvious example), give the light below a bit of that color to suggest that the light has bounced up from the ground.

Default Lights

If there are no lights placed in the scene yet, why is the scene illuminated? Until you place your own lights, there are two unseen light sources at work, at diagonal corners of the model, with one placed above the model and the other placed below. When you place a light into the scene, these two default lights turn off.

Place Omni Lights

1. Make the Top view active, and use the Zoom All tool to zoom out in all views so that the model takes up about half of each view (the Camera view will not change)

2. In the Create panel, choose the Lights category, then choose the Omni button

An Omni light is like a bare bulb that casts light an infinite distance, in all directions. It can be set to cast shadows or not. If shadows are off, the light is like x-rays, traveling through everything.

3. In the Top view, position the cursor at the Northwest corner of the barnyard, and click to set the shadow-caster, as shown in this image:

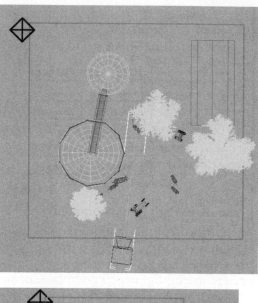

4. Position the cursor south of the model, and click to set the fill light:

Normally you would need at least one more light to illuminate the right side of things, but to keep things simple, these two lights will suffice, giving all the illumination needed for the view at Frame 0.

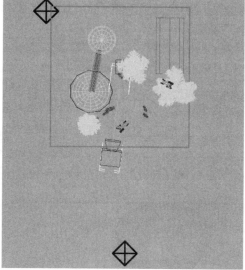

5. After placing the second light, switch to the Modify panel, and near the top of the second light's parameters, uncheck Cast Shadows:

6. Set the Light's HSV values
   to H = 35 S = 20 V = 180

General Parameters

Type: Omni

On: ☑   Exclude..
☐ Cast Shadows
R: 180   H: 35
G: 178   S: 20
B: 166   V: 180
Multiplier: 1.0

7. Select the shadow-caster light

8. In the Modify panel, set the shadow-caster's Hue,
   Saturation, Value spinners to H = 35 S = 20 V = 255

9. Make sure the Cast Shadows check box is checked

General Parameters

Type: Omni

On: ☑   Exclude..
☑ Cast Shadows
R: 255   H: 35
G: 251   S: 20
B: 235   V: 255
Multiplier: 1.0

Shadow Maps

10. Scroll to the bottom of the Modify panel
    and open the Shadow Map Params rollout

11. Set the Map Size spinner to 1000

+ Attenuation Parameters
+ Shadow Parameters
– Shadow Map Params

Bias: 1.0   Size: 1000
Sample Range: 4.0
☐ Absolute Map Bias

+ Atmospheres & Effects

Map Size is measured in number of pixels. To understand Map Size, use a mental image: think of a Shadow Map as if the renderer looks down upon the scene, calculates where shadows would fall, creates a bitmap of just the shadow shapes, and drapes that bitmap down over the scene at rendering time. With Shadow Map Size set to 1000, if you could open the shadow bitmap in Photoshop, it would measure 1000 x 1000 pixels. That's a moderately large bitmap, which would produce well-defined shadows. If the Map Size is too low, the shadows will get grainy, particularly at the edges.

Ray Traced Shadows

There is an alternative to Shadow Maps. Ray Traced Shadows are calculated more exactly than Shadow Maps. They are more accurate, and can produce shadowing from small objects like window mullions much better than Shadow Maps can. They also take much longer to render, use more system resources, and produce very crisp, hard-edged shadows. They cannot produce soft-edged subtle shadows in a softly-lit environment. They can, however, accurately portray the shadows thrown by translucent and transparent objects, which Shadow Maps cannot do well. Later chapters explore shadow issues in greater depth.

Move the Shadow-caster Up

Both lights are positioned at ground level. Move the shadow-caster up into the sky.

1. Make the Front view active, and Zoom Extents, to see both lights

2. Select the shadow-caster Omni Light

3. Choose the Move tool, then right-click over the Move tool to bring up the Transform Type-In

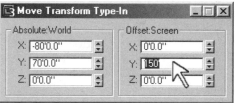

4. In the Transform Type-In, type 150′ in the Offset:Screen Y: spinner and hit Enter on the keyboard. The Omni Light will rise 150′ up. Close the Transform Type-In

5. Make the Camera view active, and render the view

Shadows help the image tremendously. Note that in this image a couple items (tractors and cows) have been moved slightly to improve the image. There is one more thing you should try before putting the barnyard away. The ground plane in this image is dull, and there is nice grass for a ground plane in the background image. If you just delete the ground plane, there will be no more shadows.

The answer is to apply a material to the ground plane that is totally transparent, but accepts shadows.

Matte / Shadow Material

1. Select the Barnyard Base

You may have noticed in the last rendering that the Base isn't quite wide enough- the Barn's shadow was falling off the edge of the Base. Make the Base wider:

2. In the Modify panel, set the Width of the Barnyard Base to 300'

3. Open the Material Editor, located just left of the Render teapots

4. In the Material Editor, select the button on the right that is labeled Type, and is currently set to Standard ———

5. In the Material / Map Browser, choose Matte Shadow, choose OK

6. In the Material Editor, check the check box labeled Receive Shadows

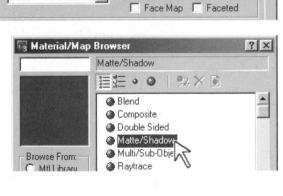

7. Choose the Assign Material To Selection button (make sure you still have Barnyard Base selected), then close the Material Editor

9. Render the Camera view

The scale of the grass is way off, but that is a topic for a later chapter.

The most sophisticated objects in this scene are the trees and the railing. All the other objects, even the cows and tractors, are just boxes and cylinders and such, and they don't even have materials assigned. It's a crude model, but it is presented with some thought to composition and story-telling elements, and it's not so bad to look at. Light, shadow, and view composition do more toward making an image compelling than ultrarealistic modeling and materials do.

Summary

While a scene made from Standard Primitives is far from realistic, Primitives are perfect for making architectural massing models. They are parametric, so you can test the dimensions of building components quickly and interactively, even dragging spinners to watch forms grow. The Extended Primitives were not used in this chapter, but if you make massing models with VIZ, you'll want to explore those.

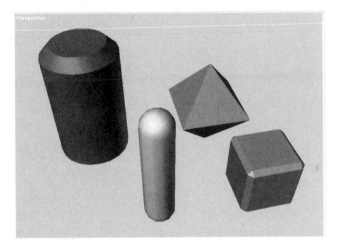

This chapter's activities centered around creating objects of various types, manipulating their creation parameters, and applying Transforms (Move, Rotate, Scale) in various ways to the objects. If you are an experienced CAD user, but have never used a program that is completely 3D oriented, as opposed to being a drafting/modeling package, the methods for the simple tasks of creating objects and changing their positions probably seem at least novel to you, and maybe even somewhat disconcerting. Things are done a little more "freestyle" in VIZ than what you are used to, and that's ok. You'll have to accept that complete accuracy in VIZ is neither a realistic aim nor a particularly useful one. When you begin to build a scene in VIZ, put away your drafter's hat, and think more of traditional model-building. Is a model built of basswood, wire, Plexiglas, and chipboard accurate? If it could be built with great accuracy, would anyone like it better- would anyone notice? And perhaps most importantly, would anyone be willing to pay more for such an accurate model? Well there's really not much difference between building a presentation model from sticks and chipboard, and building one with a computer program. If you find yourself stuggling to move a corner vertex of a wall an eighth of an inch to make a perfect fit with the floor structure below, think of a saying popular among carpenters: "we ain't buildin' no piano". If it looks good, it is good.

86

CHAPTER 3

AEC Objects: The Tower

In this exercise you will build a model of a tower and create an animation of the camera winding up the tower's spiral stairs. Nearly all of the parts of this model are parametric objects, built by assigning values to various parameters, such as the tread size, riser height, and railing height of a spiral stair. You will be amazed at how quickly and effortlessly the basic elements of a building can be defined and assembled.

In the first chapter you assigned materials from the pre-built VIZ material libraries. In this chapter you'll go a step further, loading materials from the libraries into the Material Editor and making changes to the materials before assigning them to objects.

This chapter focuses on:

- VIZ's parametric architectural objects such as Wall, Door Window, Stair, and Railing
- Movement along a path
- Shape editing
- Mesh Editing
- The Material Editor
- Lighting

Setup

1. Start VIZ, or choose File / Reset. Since you saved VIZstart.max in the first chapter, your new scene should open using feet and decimal inches.

2. You need four views showing; Top, Front, Left, and Perspective

Wall Object

1. Make the Top view active

2. From the Create panel, choose the second Category button, Shapes

3. From the various shapes, choose Rectangle

4. Near the bottom of the Create panel, open the Keyboard Entry rollout

5. Enter a Length and a Width of 28′ each, and choose Create

6. Right-click anywhere over the Top view to turn off rectangle creation mode

7. Select the Grid button at the Status Line to turn off the Grid in any views displaying one

8. In the Create panel, switch to the Geometry category button

9. Open the subcategory drop-down list, and choose AEC Extended

10. Choose Object Type Wall

11. In the Parameters rollout, enter:
    Width = 1'
    Height = 60'
    Justification = Left

12. Open the Keyboard Entry rollout

13. Choose Pick Spline

14. Position the cursor over the rectangle (so it shows a white cross), and click. A tall box will appear

15. Zoom Extents All

16. Make the Perspective view active, and Arc-Rotate Selected to get a good aerial view of the walls, then right-click to stop Arc-Rotating

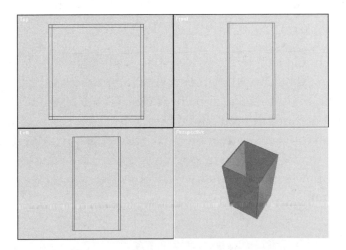

You can draw walls without drawing a 2D shape first, but that would require setting up a grid and snapping to it to draw segments of correct lengths, and in VIZ, working off a grid is rather inefficient and a bit difficult. Presumably, the floor plan exists as 2D CAD in a drafting program. If that drafting program has the ability to automatically draw centerlines for all wall segments, then you're in good shape- put the centerlines on their own layer, import or link that layer, choose the Wall tool, set Justification to Center, turn on Pick Spline, and choose any of the centerlines to create the walls. If your drafting program doesn't generate centerlines automatically, then you will either have to draw centerlines yourself, or isolate one side or the other of all walls onto their own layer.

Wall Object Terminology

Vertex: Points at the bottom of the wall that define the wall's shape in plan.

Segment: A length of wall between two vertices. You can select any Segment and change its height or width (thickness) parametrically.

Profile: The shape of a segment in elevation. ——— You can reshape any Profile of a wall, inserting new Profile vertices along the top or bottom of the Profile, and moving those vertices to add gables to the Profile.

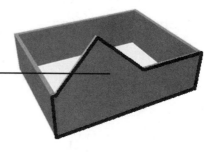

Curved Walls

In VIZ 3, Wall objects are limited to straight segments. You can use an arc as the basis for a Wall object, but when it comes time to cut openings into the curved wall for parametric Door and Window objects, things will fall apart. If your design includes curved wall segments, you should isolate the arcs of those wall segments on their own layer in your drafting program. The curved segments should be drafted as closed loops, meaning that both sides of the wall should be shown, as well as endcaps. In VIZ, you'll extrude those shapes to the height of the wall, build boxes the size of door and window openings, and use Boolean tools to subtract the boxes from the curved wall segment. It is certainly more work than the automatic openings created as you insert VIZ's parametric Door and Window objects into parametric Wall objects. Hopefully future releases of VIZ will support parametric curved walls.

Creating Gables

1. Make the Perspective view active, then type W to maximize it

2. Type the F3 function key to set the view mode to Wireframe

3. Switch to the Modify panel

4. Turn on the Sub-Object button, then switch to Sub-Object Profile

5. In the Perspective view, pick on any of the four wall segments (a grid will appear)

6. Set the Gable Height spinner to 15', then choose Create Gable

7. Select another wall segment, and give it a 15' gable. Repeat until all four Wall segments have gables

8. Turn off Sub-Object

9. Save the scene as Tower.max, in the C:\Viztutorials\Chapter3 folder

10. From the menus, choose Tools / Floaters / Display Floater

11. In the Display Floater, choose Off / By Name, and turn off Rectangle01. Close the Display Floater

Door Object

1. Switch to the Create panel

2. At the Status Line, select the Snap button, then right-click over it.

3. In the Grid and Snap Settings dialog box, choose the Clear All button, then check Midpoint and Endpoint. Close the dialog box

4. In the Create panel, choose the Geometry category button, open the subcategory drop-down list, and choose Doors

5. Choose Pivot

6. In the Perspective view, position the cursor at the bottom of the right-hand front wall, near the middle, as shown in the image at right. The edge will highlight, and you will see a square marker indicating that Midpoint snap is active

7. With the Midpoint marker visible, click-hold, then drag to the right along the bottom edge. As you near the corner, the Endpoint snap should take effect. When you see the Endpoint snap active, release the mouse, to set the width of the door

8. You now need to set the depth of the door. Position the cursor at the *top* of the wall you are inserting the door into, and on the *inside* edge of the wall, (point #3, shown at right). With the Endpoint snap marker showing, click to set the depth of the door

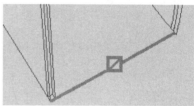

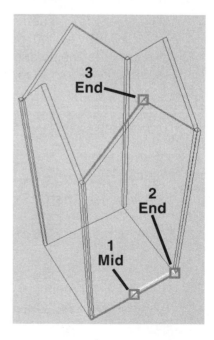

9. Now you need to set the height of the door. Move the cursor completely off the walls, so no snaps can take effect. Move the cursor upwards until the door begins to grow some positive height. Click to set any positive height.

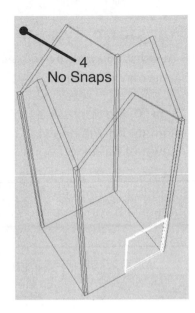

10. Immediately address the door's parameters in the Create panel:

    Set the Height and the Width to 8′

    Leave the Depth set to 1′

    (If the door does not respond, switch to the Modify panel to change the parameters)

If you set the view to a shaded display mode (F3 key), you can see that the door object has cut an opening for itself in the wall object. When you move the door, the opening in the wall will follow. If you clone the door, openings will be cut for the clones as well. The only time the automatic carving of the door openings will not occur is if you create a door object in one wall plane and try to move it or clone it to another wall plane.

As you were establishing the depth of the door in the preceding steps, you snapped to a point on the inside at the top of the wall to set the depth. The proper point on an inside corner at the bottom of the wall would have worked just as well, but sometimes picking the proper point at a crowded corner can be difficult. VIZ snaps are not as good as they could be at finding the correct snap point in a group of clustered snap points, and remember that you cannot snap to points on faces that face away from your point of view (backfaces), which adds to the confusion. When using snaps to insert doors and windows, remove the possibility of snapping to the wrong point by finding a less cluttered, unambiguous snap point elsewhere on the geometry.

Transform Type-In

The door needs to be moved 4′ to the left to center it in the wall. You'll do this with a Transform Type-in.

1. With the door still selected, choose the Move tool, then right-click over the Move tool. The Transform Type-in dialog box appears

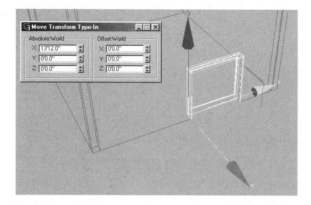

The Transform Type-in offers a means of moving, rotating, or scaling things accurately, without having to use snaps. There are two sets of X,Y, and Z spinners – the left-hand set for Absolute transforms (transforms relative to the scene's 0,0 point), and the right-hand set for Offset transforms (transforms relative to the object's current position/ orientation). In most cases, you will want to use the Offset spinners on the right.

2. Look at the Transform Gizmo, and determine the axis in which you need to move the door. In the image above, it is the Y axis (but it may be the X in your scene)

3. In the Move Transform Type-In dialog, enter –4′ (note the minus to send the door left) in the appropriate (X or Y) axis spinner on the right (Offset:World), then hit the Enter key on your keyboard. The door should move left 4′, and be centered in the wall

4. Close the Transform Type-In

Door Parameters

In the Modify panel, adjust the parameters of the door:

1. Zoom Extents Selected to get a close-up view of the door

2. Check Double Doors

3. Check Flip Swing

4. Use the Open spinner to open the door 45 degrees

5. Set the Frame Width to 6″

6. Set the Frame Depth to 2″

7. Check Generate Mapping Coordinates

8. Set # Panels Horiz: to 3

9. Set # Panels Vert: to 4

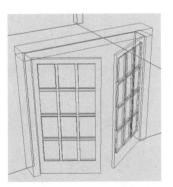

10. Zoom Extents to see the whole tower, then Arc-Rotate if necessary, to get a good view of the other front wall

11. Using the same setup and techniques that you used for the first door, create a second one, centered in the other front wall

13. Save the scene

Window Object

1. Zoom and Pan until you have a good close view of the top of the tower, as shown here:

2. In the Create panel, open the subcategory drop-down list, and choose Windows

3. Choose the button labeled Sliding

You are about to make double-hung windows. A sliding window in a vertical configuration is a double-hung.

4. Position the cursor at the lower-left corner of the gable. When the snap marker shows, click-hold, and drag to the lower-right corner of the gable, and release to set the width of the window. Move the cursor to the peak of the gable, and pick the inside corner of the peak to set the depth of the window.

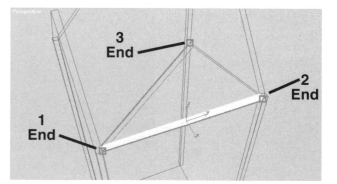

5. Move the cursor upward and away from the wall, so that no snaps are active, and when the window has positive height, click to set any positive height

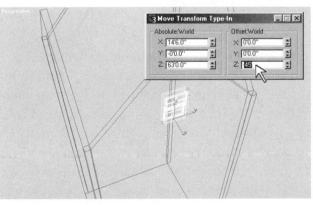

You now have a window the width of the wall.

6. Immediately address the parameters of the window:

> Set the Height to 6′
>
> Set the Width to 4′
>
> Set Frame Horiz. Width to 4″
>
> Set Frame Vert. Width to 4″
>
> Set # Panels Horiz. to 2
>
> Set # Panels Vert. to 2
>
> Check Generate Mapping Coords

Creating the window the width of the entire wall segment, then setting the width to the proper dimension is the fastest way to center a window in a wall.

7. Turn off snaps, choose the Move tool, then right-click over the Move tool to show the Transform Type-in box.

8. In the Offset: World Z spinner, type −45′, then hit Enter on your keyboard. The window will drop 45 feet. Close the Transform Type-In

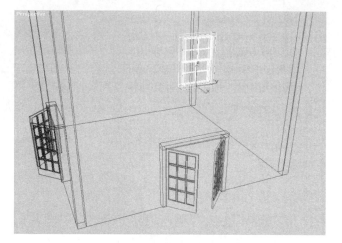

9. Zoom Extents. The window should be a few feet above the door, and centered

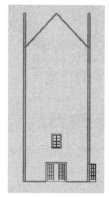

10. Type W to return to four views, then make the view active which gives you an orthogonal view of the window (either the Front or the Left view). If needed, Zoom Extents to see the entire building

Array

1. With the window still selected, choose the Array tool from the toolbar at the left of the interface.

The Array dialog box is a bit intimidating at first glance, especially if you are used to AutoCAD's array command, which is quite simple. The first settings to address are in the Array Dimensions group at the bottom: choose to make just a single row or column of objects (a 1D array), or multiple rows and columns (2D array), or a 3D array, with rows, columns, and levels. The number of objects in the array is entered in the Count spinner. The Array Transformation group is for entering the distance between objects for a linear array, or the angle between objects for a circular array, or a scaling percentage for a scaled array (such as arraying a single circle outward to draw a bulls-eye).

98

2. In the Array dialog box:

   Set the Incremental:Move:Y spinner to 12'

   Set the 1D Count spinner to 5

3. Choose OK

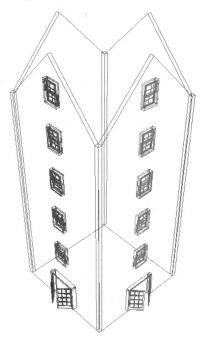

After a momentary pause, you should have windows arrayed up the wall

4. Use the same techniques and procedures to add four windows to the other front wall (start with step 1 under Window Object, page 94). If you are feeling energetic, add windows all around. Feel free to add other types of windows, and vary the parameters

5. When all the windows are in, save the scene

Stair Object

Modeling stairs from scratch is one of the more tedious tasks in architectural modeling with computers. VIZ's parametric Stair object removes that tedium; the amount of control you have over the configuration of the stairs makes them one of the more enjoyable things to model. Like all parametric objects, the best approach to creation is to just drag out a Stair object of any size and orientation, and then move the stair around and edit the properties in the Modify panel to fit the stair perfectly to the model. When the design of the building changes, updating the stair to fit the new design will take only a minute or so.

Add a Spiral Stair to the Tower model:

1. With four views showing, Zoom Extents All to see the entire tower in all four views

2. Make the Top view active

3. Select the Grid button to show the Grid

4. Make the Snap button active, then right-click over the Snap button, and set Snaps to Grid Points only

5. In the Create panel, in the Geometry category, open the subcategory drop-down and choose Stairs

6. Choose the Spiral Stair button

7. In the Top view, click-hold at the grid intersection in the center of the tower, drag out until the stairs appear. Release the mouse button to set a radius (any radius will do-you will adjust the parameters later)

8. Move the cursor upwards, until the stair takes on a positive height in the Front view. Click to set any positive height

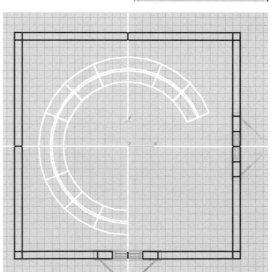

9. Turn off Snap and Grid

10. Address the parameters of the Stair:

Check the box for Stringers

Uncheck the box for Carriage

Check the box for Center Pole

Check the boxes for Rail Path Inside, and
Rail Path Outside

In the Layout field, set Radius to 10′

Set Revs to 3

Set Width to 5′

In the group labeled Rise, each spinner has a pushpin button to the left of it. You cannot change the value for whichever spinner is pinned. You want to change the Overall value, which by default is pinned, so pin a different spinner:

In the Rise group, pushpin Riser Ct.
Set the Overall spinner to 60′.
Pushpin the Overall spinner.
Set the Riser Ct to 80 (this sets the Riser Ht to 9″ risers).

Check Generate Mapping Coordinates.

Open the Strigers rollout.
Set the Depth spinner to 1′6″.
Set the Width to 2″.
Set the Offset to 6″.

Open the Center Pole rollout.
Set the Radius to 5″.
Check the box for Height.
Set the Height spinner to 64′.

Open the Railings rollout.
Set the Height to 0.
Leave the Offset set to 2″.

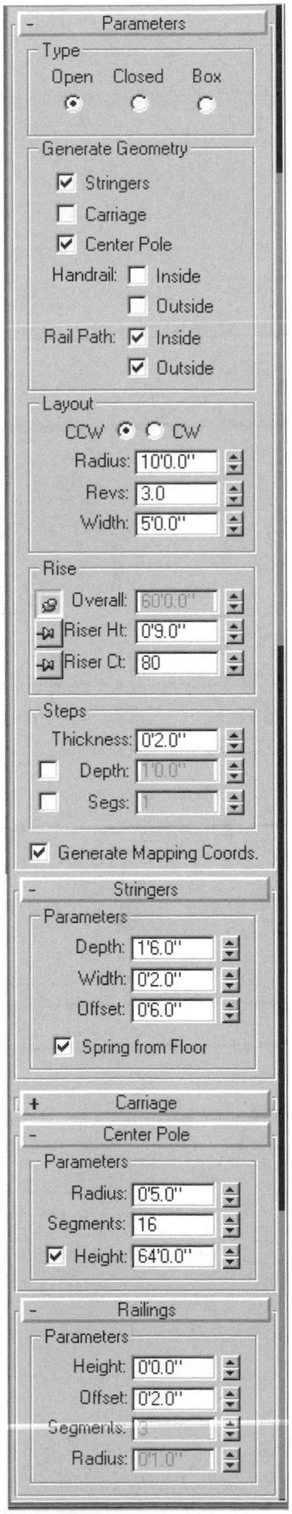

Your Stairs should look like this:

You will use the Railing Paths to do two things: they will be the paths for Railing objects, so you can quickly add handrail, posts, and ballusters, and later you'll make a Target Camera and tell the Camera to move up the outside Rail Path, and the Target to move up the inside Rail Path.

11. Save the scene

Stair Terminology

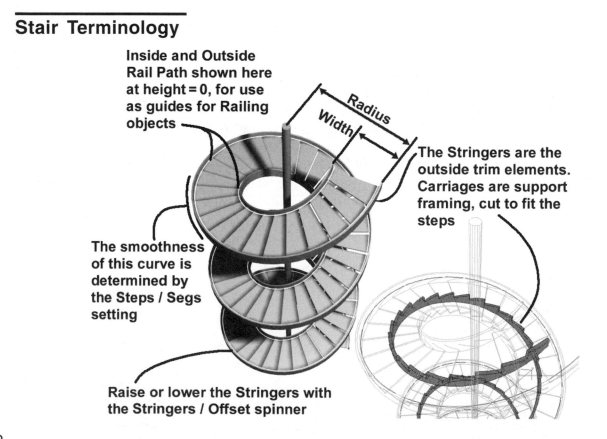

Inside and Outside Rail Path shown here at height = 0, for use as guides for Railing objects

Radius

Width

The Stringers are the outside trim elements. Carriages are support framing, cut to fit the steps

The smoothness of this curve is determined by the Steps / Segs setting

Raise or lower the Stringers with the Stringers / Offset spinner

The Display Floater

Since you'll be working just on the Stairs for a while, hide the other objects in the scene.

The Display Floater floating toolbox provides the fastest way to hide and show objects in the views, lock and unlock objects (to prevent or allow editing), change the display characteristics of individual objects, and show or hide entire categories of objects. The Display Floater duplicates most of the functionality of the Display panel, and is more readily accessible than the Display panel.

1. From the menus, choose Tools / Floaters / Display Floater

2. In the Display Floater, choose Off / by Name

The Turn Off by Name dialog box shows a list of all objects in the scene. Below that list are three buttons labeled All, None, Invert. These are quick-selection buttons: All highlights every entry in the list, None unhighlights all entries, and Invert inverts the current highlighting.

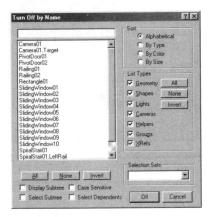

At the right of the list of objects are groups for filtering what appears in the object list, and in what order the list is arranged. For example, If you wanted all Lights in the scene to group together in the list, you would choose Sort / By Type.

The List Types group has check boxes that allow you to add or remove categories of things from the list of objects at left. To the right of those check boxes are three buttons– All, None, Invert– that give quick control over which check boxes are checked, and which are not.

Suppose you wanted to turn off the display of all Direct Lights in a scene (these lights would still cast light at rendering time, but they wouldn't have icons visible in the views). In the Turn Off by Name dialog box, you would start by choosing the None button in the List Types group. This would remove everything from the list at left. Then you would check the Lights check box, and all lights would be added to the list at left. With the Ctrl key held, you would highlight any light name containing the word Direct, then finally you would choose the Off button to hide the Direct Lights and close the dialog box.

3. In the Turn Off by Name dialog box, choose the button below the list of objects, labeled All

4. Hold the Ctrl key on your keyboard, and choose SpiralStair01, SpiralStair01.LeftRail, and SpiralStair01.RightRail from the list, to unhighlight them

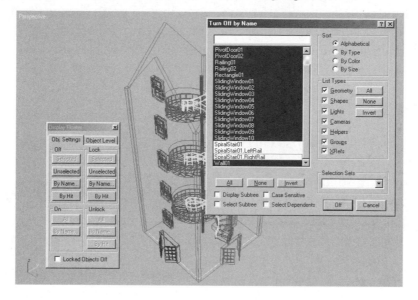

5. Choose the Off button. Only the Stairs and the two Rail Paths should show in the views

6. Move the Display Floater to a place where it is out of the way, but readily available. The best place might be over the MAXScript Mini Listener, which are two small windows, one pink and one white, at the left side of the Status Line. You won't make frequent use of the Mini Listener, and the Display Floater fits neatly over it

You will get a chance to use some other features of the Display Floater in later chapters.

Railing Object

You were introduced to the Railing object in Chapter 2, using it to make a fence around the grain pile. Now you'll use it for the handrail, balusters, and posts for the Spiral Stair.

1. Make the Perspective view active, type W to maximize the Perspective view, then Zoom tight to the stairs. Arc-Rotate to get a good view of the stairs

2. In the Create panel and the Geometry Category, open the subcategory drop-down list, and choose AEC Extended, then choose the Railing button

3. Choose the Pick Railing Path button

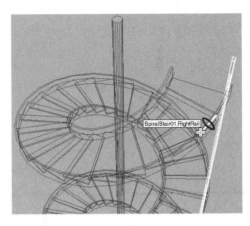

4. Position the cursor over the outside stringer of the stairs, where the outside rail path (which is set to elevation zero) is buried in the geometry of the stairs, and when you see the cursor show an icon of a vector and a ring, as shown here, click

A Railing system appears. Since it only has one segment, it looks like a pole.

In a situation such as this, when you need to pick on an object obscured by other objects in a cluttered view, the safe way to make the correct pick is to type H instead of picking directly in the view. Because you are in a Pick Object mode, typing H will open a Pick Object dialog box, from which you can pick exactly the object you want.

5. Change the parameters of the Railing:

   Set the Segments spinner to 16

   Set the Top Rail Profile to Round

   Set the Top Rail Depth and Width to 2.5"

   Set the Top Rail Height to 3'6"

In the Lower Rail(s) group, leave the Profile set to Square

Set the Lower Rail Depth and Width to 2″

Check Generate Mapping Coordinates

Open the Posts rollout

Set the Profile to Round

Set the Posts Depth and Width to 4″

Choose the Post Spacing button

In the Post Spacing dialog box,
set the Count to 12, then close the dialog box

Open the Fencing rollout

Leave the fencing Type set to Pickets

Set the Fencing Depth and Width to 2″

Choose the Picket Spacing button

In the Picket Spacing dialog box, set the
Count to 8, then close the dialog box

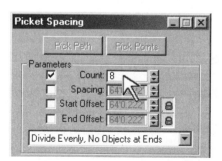

The Fencing Type drop-down list shows the choices None, Pickets, and Solid Fill. Pickets is synonymous with balusters. Solid Fill is for Plexiglas rail systems, for example.

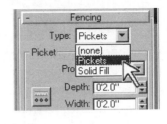

6. Render the view

7. Repeat the Railing creation process to make a similar Railing for the inside of the Stairs

8. Since this rail doesn't have as far to travel as the outside one, give the inside rail 12 segments, and set the Post count to 8

9. Save the scene

Camera Paths

Instructing a Camera to follow a path is very easy in VIZ 3, and making a Camera climb Stairs is also easy because the two Rail Paths that are part of the Stair object make perfect paths for the Camera. The Rail Paths for the Stairs you've just created are at elevation zero, so that the Railing objects could be built upon them and sit on the Stair treads. You'll copy the two paths up 5'6", create a Camera, and link it to the paths:

1. Type W to restore four views, then make the Front view active

2. Zoom Extents

3. Type H, select SpiralStair01.RightRail

4. From the menus, choose Edit / Clone

5. In the Clone Options dialog box, choose Instance as the clone type, name the new path Camera Path, and choose OK

Why choose Instance? Suppose you decide later that the Stairs are too large at Radius 10', and you change the Radius to 8'. Because the first set of Rail Paths are part of the Stair object, they'll change to an 8' radius as well. If the second set of paths are cloned as Instances, they will also change radius, and the Camera will move inward accordingly to remain over the Stairs.

6. Choose the Move tool, right-click over it, and in the Transform Type-In, in the Offset:Screen Y: spinner, type 5'6", then hit Enter. Leave the Transform Type-In open

7. Choose the Select by Name button (normally you can just type H to bring up the Select by Name dialog box, but currently your cursor is active in a spinner in the Transform Type-In, so keyboard shortcuts won't work)

8. In the Select Objects dialog box, select SpiralStair01.LeftRail

9. From the menus, choose Edit / Clone

10. In the Clone Options dialog box, choose Instance as the clone type, name the new path Target Path, and choose OK

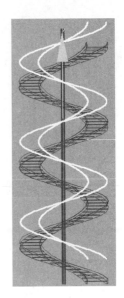

11. In the Transform Type-In, in the Offset:Screen Y: spinner, enter 5′6″, then hit Enter. Close the Transform Type-In

You should now have two new helix paths 5′6″ above the two Rail Paths. Now make a Camera.

Follow Path

1. Make the Top view active

2. In the Create panel, choose the Cameras button, then choose Target

3. Drag out a Target Camera anywhere in the Top view

4. In the Camera's parameters, choose the 20mm Stock Lenses button

5. From the menus at the top of the interface, choose Animation / Follow Path

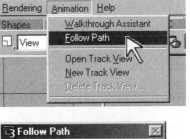

6. The Follow Path dialog box appears. Choose OK to close the dialog box

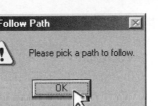

7. In the Front view, click on the Camera Path

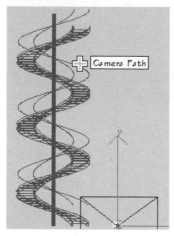

The Camera will jump to the start of the outside helix path.

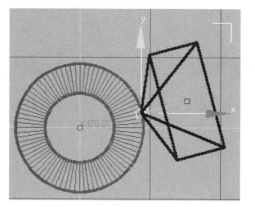

Now repeat the Follow Path procedure with the Camera Target:

1. Type H and select Camera01.Target

2. From the menus, choose Animation / Follow Path

3. Choose OK to close the Follow Path dialog box

4. Pick on the Target Path in the Front view

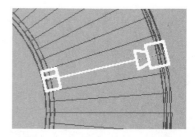

The Camera Target will jump to the start of the inside helix.

5. Make the Perspective view active and type C to change it to the Camera view

6. Drag the Time Slider below the views. You should have an animation of the Camera winding up the Spiral Stairs

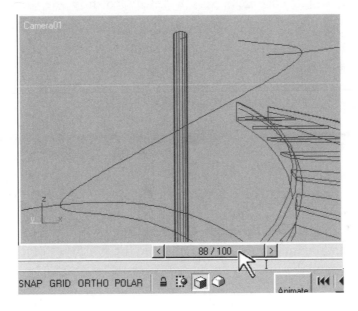

7. Save the scene

The Animation/Follow Path tool is a great shortcut for the process of assigning a Path Controller. A Controller is anything that controls animation of an object. Controllers are assigned in a few places in the interface, including the Motion panel and in Track View. Besides paths, sound can be a Controller (so that doors open when a bell chimes, for example), a surface can be a Controller (to force a car to follow a road over hills), you can assign a Look At Controller to keep an object pointed at another object (so that a cutout person or tree always faces the Camera), and there are numerous mathematical Controllers to alter an object's behavior at any point in time and space.

Render a Preview

A Preview is a rough version of the animation. It plays back in a small window, shows materials only as shades of color, does not have shadows, and does not use antialiasing (the smoothing out of jagged edges). It is just meant to show the sequence of action and give an idea of the timing and pace of the animation.

1. Make the Camera view active

2. From the menus, choose Rendering / Make Preview

3. In the Make Preview dialog box, leave all settings at their defaults, and choose Create

4. The Video Compression dialog box may appear – if so, just choose OK

The renderer will render a rough version of each frame of animation, compile those frames into an animated image (an .AVI), and open the Windows Media Player to view the Preview. Rendering the Preview should only take a few minutes.

4. When Media Player opens, the animation should play automatically. If not, choose the Play buton

5. After viewing the Preview, close Media Player

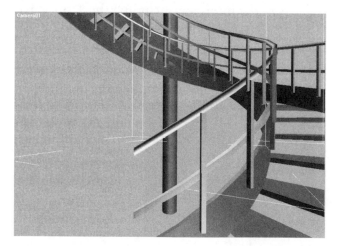

Modeling the Roof

Since 3D Studio VIZ does not have parametric roofs, it's time to do some building "by hand". You will work a flat plane into the roof planes by dividing the plane and moving vertices of the plane in the Z direction. Editing the individual faces, edges, and vertices of an object is called Mesh Editing.

1. If the Display Floater is still tucked away in the lower-left of the screen, drag it out. If you closed it earlier, choose (from the menus) Tools / Floaters / Display Floater

2. In the Display Floater, choose Off / By Name

3. In the Turn Off by Name dialog box, choose the All button below the list of scene objects, then choose Off

4. Move the Display Floater out of the way, or close it

5. Make the Top View active

6. In the Create panel, Shapes category, choose Rectangle

7. Open the Keyboard Entry rollout, set the Length and the Width to 32′, and choose Create

8. Name the Rectangle Roof

9. Make the Camera view active, then change it to a Perspective view by typing P

10. Zoom Extents All, then Arc-Rotate and Zoom in the Perspective view to get a good view of the Rectangle

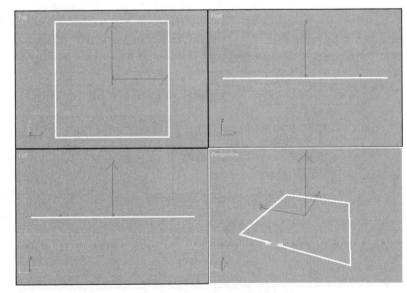

Modifiers

In the previous chapter there was a brief discussion of what a Transform and a Modifier are. The Transforms are Move, Rotate, and Scale, and a Transform changes an object's position, orientation, or size, but it doesn't change its structure. A Modifier does change the object's structure. You can add as many Modifiers to an object as you want. If you add several Modifiers to an object, perhaps to Bend, then Twist, then Taper it, those Modifiers are stacked on top of each other in the object's history, and can be accessed at any time and edited, to reshape the object.

A slide projector makes a good metaphor for Modifiers - not a modern carousel type, but one of the old projectors that you advance with a sliding bracket. Suppose you project a picture of a pig onto a wall. Then you add a slide of a pair of sunglasses in front of the pig slide: now your pig has shades. Add a slide of a cigar to the stack of slides, and the pig smokes. You are free to take the cigar slide back out of the stack, if you don't like it, and maybe replace it with a slide of earrings. The additional slides are Modifiers. You are not altering the picture of the pig, you're just adding Modifiers to it. Each modifier remains a discrete part of the history of the object, and can be edited parametrically at any time.

In VIZ, you actually can alter the original object if you want. You can create a six-sided Cylinder, apply a Taper Modifier, then a Twist, then a Bend, and if you decide at that point that the Cylinder needs more sides, you can access the original Cylinder parameters (called Creation Parameters) and set the Sides spinner to 12. The change will be sent up the stack, and the tapered, twisted, bent object will take on more definition. There are limitations to this ability. The addition of certain Modifiers can make it dangerous to edit the object's original creation parameters and the object may be ruined if you try. VIZ issues warnings about this as you access creation parameters for an object in this state.

You are going to start using Modifiers by exploring one of the most complex ones, called Edit Mesh. The Bend Modifier will be a piece of cake after you master Edit Mesh. If you don't understand what you are doing, conceptually, in this next section, don't worry about it. Mesh editing is best learned by just being walked through it a few times. You will get another chance to do some extensive mesh editing in the chapter on furnishings, and by the end of that, you will know what's going on. When you need to edit an object at its most basic levels, manipulating individual faces of the object or moving some vertices, you apply an Edit Mesh modifier to the object.

Edit Mesh Modifier

11. In the Modify panel, apply an Edit Mesh modifier to the Rectangle

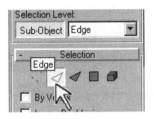

12. Select the Sub-Object button, then switch to Sub-Object Edge

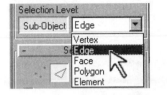

Sub-Objects are just components of an object, or a component of the Modifier itself. Many Modifiers feature Sub-Objects of various types. There are a few ways to get to Sub-Object level. There is the drop-down list you just used, and just below that drop-down list are Sub-Object shortcut buttons, that are quicker than the drop-down list.

There is also a menu accessed with a right-click that has Sub-Object tools. You will use that menu in the next chapter.

13. Show the Display Floater, and at the top of it, switch to the tab labeled Object Level

14. Uncheck Edges Only

Meshes in VIZ are composed of triangles. To improve the display, many of the edges of triangles are made invisible. Wherever two triangles can be displayed as one quadrilateral instead, the program does so. When you need access to every edge in the mesh, turn off Edges Only.

15. Select the diagonal edge (shown dashed, meaning it will be invisible when Edges Only is active)

16. Scroll to the bottom of the Modify panel, and in the Surface Properties rollout, select Visible

17. Set Snaps to Midpoint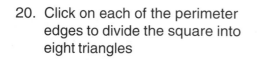

Divide Edges

18. In the Modify panel, in the Edit Geometry rollout, choose Divide

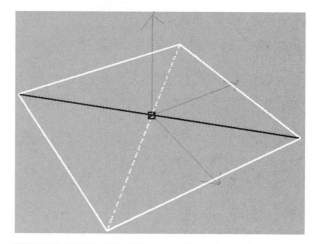

19. Place the cursor at the midpoint of the diagonal edge, and click to divide the square into four equal triangles

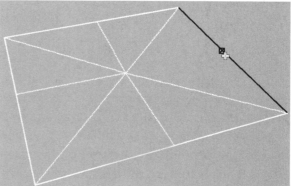

20. Click on each of the perimeter edges to divide the square into eight triangles

21. Click on the Divide button again, to turn it off

22. Turn off Snap

Extrude Edges

23. Select all the perimeter edges of the square (selected edges turn red). Click on one perimeter edge, then hold the Ctrl key and click on the rest (eight edges total)

Be careful not to move the mouse as you select edges. Move is the active command at this point, so to select without accidentally moving, the mouse needs to be held still.

24. In the Modify panel, choose the Extrude button, then enter -12 (note the minus) in the Amount spinner (do not add the inch symbol), and hit Enter on the keyboard

The various views should show the edges with a 1' thickness:

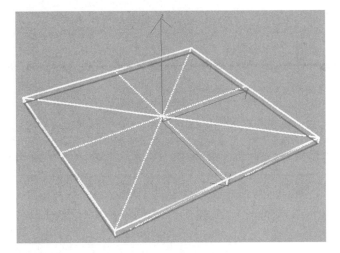

Move Vertices

1. Switch from Sub-Object Edge to Sub-Object Vertex

2. Hold the Ctrl key on the keyboard, and in the Top view drag two windows to select the vertices at the center of the square, and at the midpoints of the four edges, as shown in the following image:

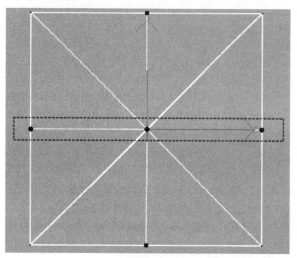

3. Make the Front view active

4. Choose the Move tool, then right-click over it to show the Transform Type-In

5. In the Offset: Screen Y: spinner, type 17, then hit Enter on the keyboard

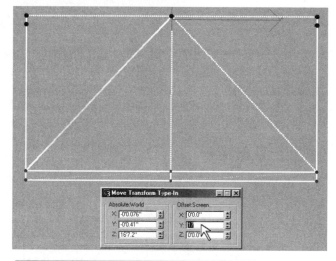

6. Close the Transform Type-In

7. Save the scene

The roof should look like this:

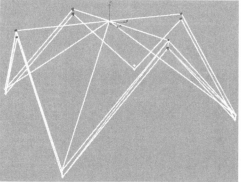

Edit Faces: Material ID Number

1. Switch to Sub-Object Face

2. Choose the Select tool

3. At the bottom of the interface, set the Window / Crossing button to Crossing Selection

⚠ If it is already set to Crossing, select it twice to reset it to Crossing. There seems to be a problem with this button- it requires resetting at times

4. In the Top view, drag a selection region around the vertex at the center of the square to select all the faces of the roof surfaces

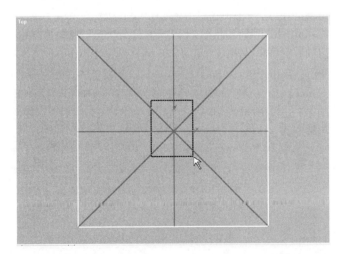

Window selection must completely surround an item to select it. Crossing selection can either surround it or just touch it. When using Window selection you will often find that as you begin to drag to make a selection of faces, one face (the face your cursor was over) will highlight red as soon as you begin to drag, and you may not want this face included in the selection. Ignore the initial highlighting; when you release the mouse, the correct selection will be made. But because of this odd behavior, you need to make sure the Select tool, and not the Select and Move tool, is active as you begin to drag a Sub-Object/Face selection. The other way to avoid this behavior is to begin to drag the selection region outside the extents of the object (not over any face of the object).

5. Scroll to the bottom of the Modify panel to see the Surface Properties rollout, and set the Material ID for the selected faces to 2

Later, when you make a Multi/Sub-Object material for the roof, material 2 will be a roofing material, and material 1 will be for the fascia.

Setting certain parts of a mesh to carry certain Material ID numbers is something you will do often. Imagine assigning materials to a model of a car. The tires could be one object, and you would design a tire material and assign it to the tires. The windshield could be another object, with its own material, and the bumpers another object, with their own material, and so on. Obviously the result would be chaotic, with dozens of objects making up the car, and a material editor cluttered with dozens of materials, all for one car. The alternative is a Multi/Sub-Object material. It is a container, basically, for any number of materials. Each of the materials nested in the Multi/Sub-Object material has an ID number assigned to it in the Material Editor. You make selections of parts of the mesh, and assign a Material ID to those faces that correspond to the appropriate material number in the Multi/Sub-Object material. The tires of the car get ID 1, the windshield ID 2, the bumpers ID 3, and so on. Once the Material ID numbers have all been assigned to the parts of the mesh, you assign a single Multi/Sub-Object material to the mesh, and all the nested materials know where on the mesh to apply themselves.

Edit Faces: Clone a Face Set

You need another set of roof planes for the ceiling. Since you've already got the roof planes selected, having just assigned Material ID 2 to them, it's simple to clone them downward a foot for the ceiling. First you need to set up so that you can drag exactly one foot downward, and that means using the Grid and Snap Settings:

1. Make the Front view active

2. Zoom Extents

3. Activate the Grid button and the Snap button, then right-click over the Snap button

4. In the Grid and Snap Settings dialog box, set Snap to Grid Points only. In the Home Grid tab, make sure Grid Spacing is set to 1 foot. Close the dialog box

5. Activate Lock Selection Set

6. Choose the Move tool

7. Place the cursor away from the Roof, hold the Shift key on the keyboard, and with the cyan-colored Grid Snap marker showing at a Grid intersection, shift-drag downward one Grid line, and release the mouse

8. In the Clone Part of Mesh dialog box, leave Clone To Element chosen, then choose OK

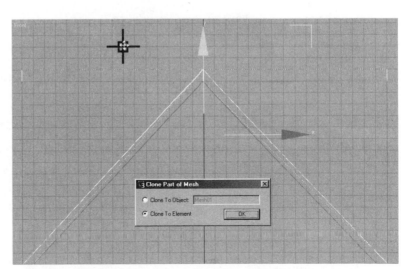

Edit Faces: Flip Normals

Face Normals, you will recall, are the sides of faces that the renderer and the viewports see. The other sides of faces, the Backfaces, are by default invisible in the viewports and in the rendering. If certain faces are facing the wrong way, there are three things you can do:

1. In the Render dialog box, indicate that both sides of every face be made renderable. This is called Force 2-Sided, and can slow rendering speed greatly.

2. In the Material Editor, make the material assigned to the problem faces a 2-Sided material. Faces in the scene assigned that material will render on both sides.

3. Select the problem faces and flip their Normals. If there are just a few faces in the scene with Normals the wrong way, this is the best solution, and this is what you will do with the ceiling planes, which are currently only visible from above, and should be visible from below.

1. Turn off Snap, turn off Grid, turn off Lock Selection Set

2. Scroll to near the bottom of the Modify panel, and in the Surface Properties rollout, in the group labeled Normals, select the button labeled Flip. The ceiling planes will now face downward, as they should

Edit Faces: Material ID Number

A few minutes ago you assigned Material ID 2 to the roof planes. Material ID 1 was assigned by default to the fascias. Now, the ceiling planes need to be assigned Material ID 3:

1. In the Surface Properties rollout, in the field labeled Material, assign Material ID 3 to the ceiling planes

When you make the Multi/Sub-Object material that will be assigned to the Roof, that material will contain three submaterials.

The modeling of the Roof is complete. All that is left to do is to raise it to the proper elevation.

1. Scroll back up the Modify panel, and click on the Face Sub-Object button to leave Sub-Object mode

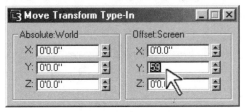

2. Make sure the Front view is active

3. Choose the Move tool, then right-click over it to show the Transform Type-In

4. In the Offset: Screen Y: spinner, enter 59, then hit the Enter key on the keyboard, then close the Transform Type-In

5. Access the Display Floater, switch to the Obj. Settings tab, choose On / By Name, and turn on Wall01

6. Zoom Extents to see the Wall and Roof

The roof should now sit properly on top of the Walls.

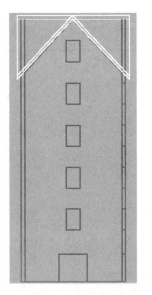

7. Save the scene.

Modeling the Landing: Spline Editing

The upper landing, at the top of the Spiral Stairs, is simple; it is a square with a half-donut subtracted from it.

The only work you've done with 2D Shapes so far has been to create a few Circles and Rectangles. Now you will be introduced to a few basic ideas about Spline Editing. The Shapes category contains two subcategories: Splines and NURBS Curves. NURBS topics are not covered in this book. A Spline is defined in the online Help as "a type of curve that is interpolated between two endpoints and two or more tangent vectors. The term dates from 1756, and derives from a thin wood or metal strip used for drafting curves in architecture and ship design." So in its simplest form, a Spline connects two vertices, can be straight or curved, and its curvature can be controlled by manipulating control vectors at each vertex.

Splines have component parts (Sub-Objects), and you will need to be familiar with the terminology of those Sub-Objects. The overall Shape is referred to in certain menus as the Base Object. The Shape may consist of a single Spline, as in the curve shown above, or it may contain multiple Splines, as in the sketch of the landing at the top of the page (the landing contains two Splines). Splines are made up of Segments. A Segment connects two Vertices. The simple Spline above has only one Segment, connecting two Vertices, and the landing Shape has ten Segments (each arc has two Segments).

In this book, Splines have served as the basis for Wall objects and Railing objects (the two helix Rail Paths are Splines), and in VIZ, Splines serve as the basis for many other things. Much of the power of modeling in VIZ comes from being able to base a 3D object on one or more 2D Shapes, and then accessing and editing the 2D Shapes to resculpt the 3D object.

The landing will serve as brief introduction, and you will practice with spline editing more extensively in the next chapter on furnishings.

Create Splines

1. Make the Top view active

2. In the Create panel, Shapes category, choose Rectangle, open the Keyboard Entry rollout, enter 28' for both Length and Width, then choose Create, to create a square that just fits inside the Wall

3. Use the Display Floater to show only the Rectangle and the Spiral Stair in the views

4. Name the Rectangle Landing

5. In the Create panel, just above the buttons for the various Shapes, is a check box labeled Start New Shape. Uncheck it

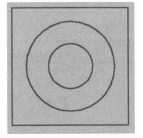

⚠ This Start New Shape check box is a bit understated in the interface, considering how important it is. You are about to draw the opening for the stairs in the landing. The opening should be part of the same shape as the rectangle. If you forget to uncheck Start New Shape, the opening will be separate from the rectangle, and you will have to join them together. Better to remember that check box before you draw.

6. In the Create panel, in the Shapes category, choose Donut

7. Open the Keyboard Entry rollout, set:

    Radius 1 = 4'8"
    Radius 2 = 10'4"
    Choose Create

8. Make the Front view active, and Zoom Extents

9. Choose the Move tool, then right-click on it

10. In the Transform Type-in, set the Offset: Y spinner to 60', then hit the Enter key to raise the landing to its correct height. Close the Transform Type-in

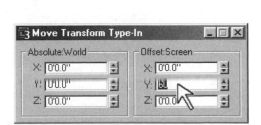

Edit Spline - Delete Segments

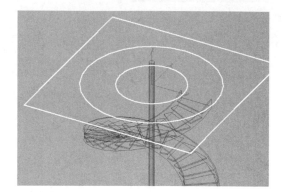

1. Make the Perspective view active, type U to change it to a User view, then Zoom Extents Selected, Arc-Rotate, and Zoom so the view looks like this: Make sure your view of the stairs matches this image, with the stairs ending at the right

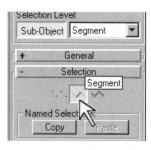

2. In the Modify panel, switch to Sub-Object Segment

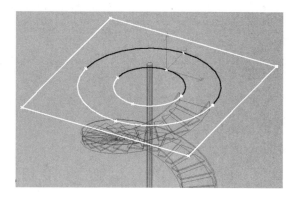

3. Hold the Ctrl key on the keyboard, and select the four Segments of the Donut that are not needed for the semicircular opening

4. Scroll near the bottom of the Modify panel and choose Delete (or the Delete key on the keyboard should also work)

Edit Spline-Connect Vertices

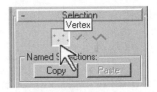

5. Switch to Sub-Object Vertex

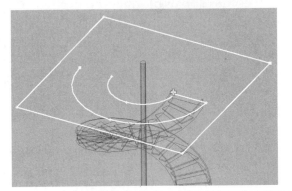

6. Scroll about halfway down the Modify panel, and choose the button labeled Connect

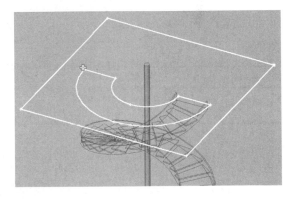

7. Drag from one end vertex of the Donut to another to cap that end, as shown here:

8. Connect the other two open end vertices by dragging between them

9. Select the Connect button again to turn off Connect mode

10. Select the Sub-Object Vertex button again to leave Sub-Object mode

Extrude Modifier

The 2D Shape is now ready to be modified into a 3D mesh by applying an Extrude Modifier.

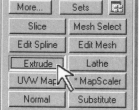

1. From the default 10 Modifiers near the top of the Modify panel, choose Extrude

⚠ Many Modifiers show no immediate effect when you apply them– their presence only becomes apparent when you begin to alter the parameters of the Modifier. Once you click a Modifier button, don't click it again if you don't see a change in the object, or you will just stack another unnecessary modifier on the object.

2. In the Extrude parameters, set the Amount spinner to – 1′ (note the minus)

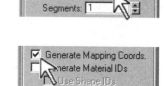

3. At the bottom of the Modify panel, check Generate Mapping Coordinates, uncheck Generate Material IDs, and leave Smooth checked (so that the curved edges at the opening render smooth)

Several Modifiers have in their parameters a check box for Generate Material IDs. When you apply a Modifier, look for this check box, and consider carefully whether you want it checked (the default is that it be checked, usually). When checked, the object is likely to carry more than one Material ID. If you want the entire object to be one material, uncheck this check box, and the whole mesh will carry Material ID 1.

4. With the User view active, type P to set it back to a Perspective view, then navigate for a good view of the landing, and Quick-Render

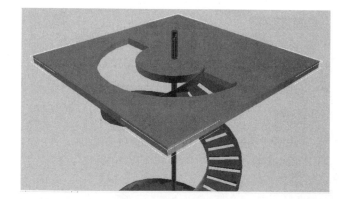

The landing should look like this:

Model the Lower Floor

The last item you will model at this point is a floor. Then you'll work on materials and lighting:

1. Make the Top view active

2. In the Create Panel, Shapes category, choose Rectangle. In the Keyboard Entry rollout, enter a Length and a Width of 28′, then choose Create

3. Name the Rectangle Floor

4. In the Modify panel, apply a UVW Map modifier to make the floor solid

5. In the UVW Map parameters, set the Length and the Width to 10′

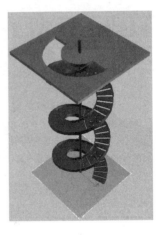

There will be many cases in which you've drawn (or imported) a 2D Shape, and you simply want to make it a solid plane. UVW Map is a good modifier for doing this because it serves double-duty; it makes Shapes solid, and it applies instructions for how images used in materials (a photo of marble, maybe) are to be applied to the geometry.

6. Save the scene

The Material Editor - Using Material Libraries

Now you will add materials to the scene. In this chapter you'll work mostly with premade materials from the VIZ material libraries. Although the first time you do a real-life visualization project, you will likely find a need for materials that are not in the VIZ libraries, these premade materials are a good way to learn the Material Editor. Rather than make the material that you need entirely from scratch, find a similar material in the VIZ libraries, and alter it to suit your needs. As you explore the premade materials and try changes, you'll start to get an intuitive sense for how the Material Editor works, and how to make realistic materials. In later chapters of this book, you will build complex, natural-looking materials, starting from scratch.

Material Editor Terminology

1. In the Display Floater, choose On / All

2. Type H, Select the walls

3. Open the Material Editor

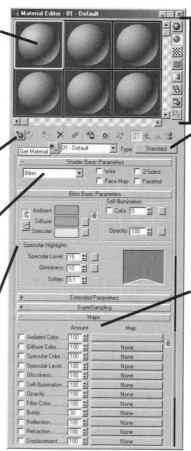

Sample Windows: This area shows 6, 15, or 24 sample swatches of materials. Samples can be spheres, cylinders, or cubes. Right-click over the active sample for options menu

Get Material button: Retrieve premade materials from libraries and from other VIZ scenes

Shader Type drop-down: Shaders are mathematics used to depict the physical qualities of a surface

Shader Basic Parameters: The qualities of the material before any mapping has been added to its appearance; what color is it in shadow, in general light, and in its highlight? How shiny is it? Is it opaque, clear, or translucent? Is it a glowing material, like neon?

Configuration Buttons: Set the Material Editor up the way you want

Material Type button: Choose a general category of material to work with: Standard, Raytrace, Blend, Multi/Sub-Object...

Maps Channels: Assign bitmap images to do a variety of tasks: give a basic pattern to the surface, make it appear bumpy, make parts of the object transparent, make the object reflective, make it shine a certain way...

Retrieve a Material From a Library

4. In the row of buttons below the six sample spheres, choose Get Material, at the left end

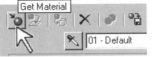

5. Near the upper-left of the Material / Map Browser, choose Browse From: Mtl Library, to access the library of premade materials that comes with VIZ

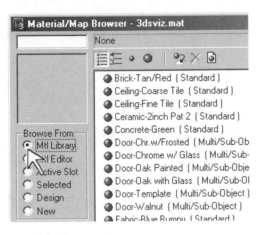

6. In the list of materials, scroll to the "S"s, and single-click on Stone-Limestne.Wall, to see a sample sphere of it.

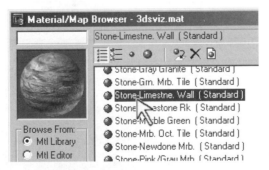

7. Move that material into the Material Editor by either double-clicking on the name, or by dragging from the name or the sample sphere to the first (upper-left) sample window in the Material Editor

8. Close the Material / Map Browser

VIZ is programmed to be very "drag-and-drop." Anytime you need to assign an item to another place in the interface, try dragging it there. You'll be surprised how much "dragability" there is.

Assign Material to Selection

Materials can be assigned two ways: either select an object and in the Material Editor choose the Assign Material to Selection button, or you can simply drag materials from the sample windows in the Material Editor, and drop them onto objects in the views. Dragging materials onto objects works well in a simple scene, but not so well in a crowded one, where you frequently drop materials onto the wrong object. A welcome addition to the Material Editor would be a button for Assign Material by Name.

1. In the row of tools below the sample windows, choose Assign Material to Selection, to assign the limestone wall material to Wall01

2. Make the Left view active, then type C to turn it into a Camera view

3. Drag the Time Slider to about frame 25

4. Quick-Render the Camera view

The limestone image is applied to the walls, but it looks wrong. The problem is in the scale of the stone, which is easily fixed.

5. Close the rendered view

Map Tiling

There are two things that control the scale of bitmaps on an object: Mapping Coordinates, and Tiling. Mapping Coordinates will be discussed shortly. For now, you'll adjust the Tiling of the image of limestone to improve its scale. Tiling is how many times the bitmap repeats over a given distance.

1. In the Material Editor, scroll to see the Maps rollout (if it is rolled shut, open it)

The Maps rollout shows numerous ways that images can be used within a material, and in this Limestone Wall material, a photo of limestone has been used to provide the basic appearance of the material (Diffuse Color Map), and also to give the material the appearance of bumpiness (Bump Map).

2. In the Maps rollout, uncheck the check box at the left of the Bump channel

3. Choose the bar labeled LIMESTO3.jpg in the Diffuse channel of the Maps rollout

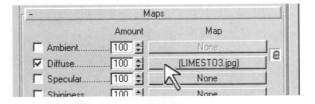

You've now dropped down a level in the Material Editor. Complex materials involve multiple levels, maybe up to half a dozen, with several sibling parameter sets at each level. While the multitiered, hierarchical organization of complex materials in the Material Editor may make your head spin at first, you'll see that complexity is often the key to natural, believable materials.

4. In the Diffuse Map level, set Tiling: U to 3, and set Tiling: V to 6

5. Quick-render the Camera view

The limestone is now a much more reasonable size. Now put the bumpiness back.

6. Choose the Go to Parent button to return to the Maps channels at the uppermost level

7. In the Maps channels, drag-and-drop the label LIMESTO3.jpg from the Diffuse Color channel over the Bump channel (replacing the version of LIMESTO3.jpg already there), and in the Copy [Instance] Map dialog box, choose Instance, so that any changes you make to the map in one channel will be mimicked in the other

8. Quick-render the Camera view again. The stone wall should now look bumpier, and therefore more realistic

Door Material

As you create Door objects and Window objects, Material ID numbers are automatically assigned to the various parts of the doors and windows. The VIZ library contains door and window materials that are Multi/Sub-Object materials, with material numbers corresponding to the Material IDs on the objects.

1. Drag the Time Slider to frame 11, so you can see doors and windows through the Camera view

2. Type H, and Select both Doors (called PivotDoor01 and PivotDoor02)

 When the Material Editor is open, typing H to open the Select Objects dialog box often does not work. In this case, use the Select by Name tool

3. In the Material Editor, make the second sample sphere active by clicking on it

4. In the row of tools below the sample spheres, choose Get Material, at the left side

5. In the Material / Map Browser, scroll to find material Door-Oak with Glass

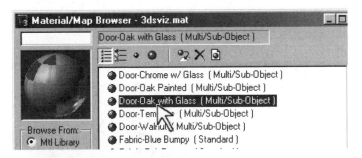

6. Drag the Door-Oak with Glass material onto the second sample window

7. Choose Assign Material to Selection

Window Material

1. From the Material / Map Browser, drag the Window-Wht Wd/Clr material onto the third sample window

2. Type H and select all windows, then assign the window material to the windows

When making multiple selections in the Select Objects dialog box, if the objects to be selected are listed together (like the list of windows), you can just click-hold on the first item, and drag over the others, then release to select them all. You only need the Ctrl key held down when selecting multiple objects that are not listed together.

Stair Material

3. Drag the Stair-Wood/Brass material onto the fourth sample window, select SpiralStair01, and assign the stair material to it

Railing Material

4. Drag the Metal-Brass material onto the fifth sample window, and assign the brass to the two railing objects

5. Quick-Render the Camera view

Background Color

6. To change the background color, choose Rendering / Environment from the menus, click on the background color swatch, select a light color, then close the Environment dialog box

7. With the Time Slider at frame 11, render the Camera view

8. Save the scene

Multi / Sub-Object Material

As you modeled the roof with mesh editing tools, you assigned Material ID numbers to various parts: ID 1 for the fascias, ID 2 for the roof planes, and ID 3 for the ceiling planes. Now you will make a Multi/Sub-Object material, consisting of three sub-materials, to assign to the roof.

1. Scroll the Time Slider to frame 100, so you can see the ceiling through the camera

2. In the Material Editor, make the last unused sample window active

3. Below the sample windows, at the right, select the button labeled Standard

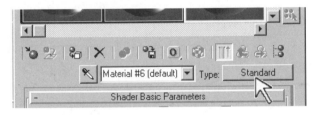

4. In the Material / Map Browser, highlight Multi / Sub-Object, then choose OK

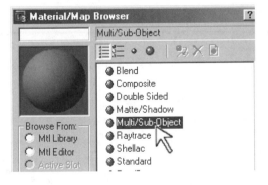

5. In the Replace Material dialog box, leave it set to Keep old material as sub-material, choose OK

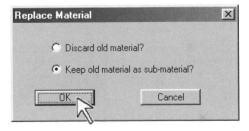

In this case the choice here is unimportant, but when you've already done some work on a Standard material, and then decide to convert to a Multi/Sub-Object, you want to keep the changes you've made as a submaterial.

6. Select the button labeled Set Number. In the Set Number of Materials dialog box, set the spinner to 3, then choose OK

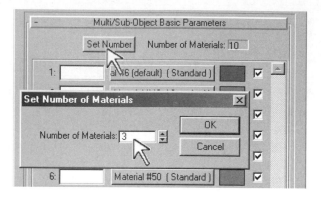

Multi / Sub-Object Naming

Naming is highly important in the Material Editor. When working with a complex material, which will have several tiers of settings, and probably a few sibling parameter sets at each tier, you need a logical and consistent naming scheme for each tier and each sibling, so you know where you are as you navigate through the material.

7. Highlight the top-level name, and rename it Roof / Ceiling

8. Name the submaterials Fascia, Roof, and Ceiling

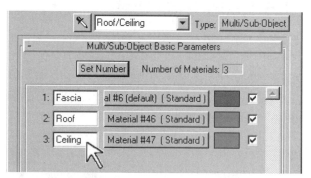

9. Choose the Get Material button, and position the Material / Map Browser and the Material Editor side-by-side

Populate Submaterials

10. In the Material / Map Browser, verify that the
    Browse From field is still set to Mtl Library.
    Then from the File field (bottom left), choose
    Open

11. In the Open Material Library dialog box, highlight
    3dsviz_big.mat, then choose Open

This is a large library, and will take several seconds
to open. If you cannot open 3dsviz_big.mat at all,
open Paint.mat instead, and substitute any color
paint you want in step 12.

12. Drag material Paint-Light Grey 1 from the material library, and drop it over the first
    of the submaterials

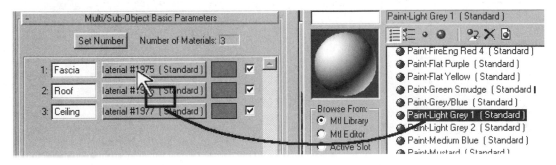

13. Open material library Roof.mat

14. Drag material Roof-Red Spanish Tile to the second submaterial bar

15. Open material library Wall.mat

16. Drag material Wall-Stucco 12 to the third submaterial bar

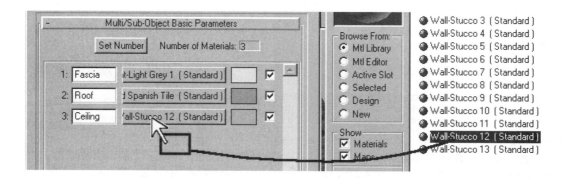

17. Close the Material / Map Browser

18. Drag the sample sphere onto the Roof to assign the material. You can drag into any view in which the roof is visible. As you're dragging, before you release the material over the object, wait for a tooltip to appear on the object, showing the object's name, to ensure you drop onto the correct object

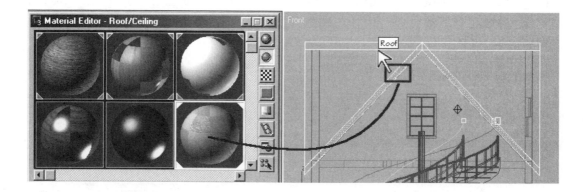

Additional Samples

You need to load another material from a library, but all six sample windows have been used. You can either switch from six larger sample windows to fifteen or twenty-four smaller ones, or you can scroll the large windows to reveal additional ones.

1. In the Material Editor, position the cursor over the division between any two sample windows, and when the cursor shows a hand, pan towards the upper-left to reveal unused sample windows

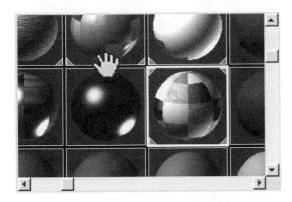

Notice that there are scrollbars at the right of and below the sample windows, which also let you pan to see other spheres.

2. Right-click over the active window, and in the pop-up menu, choose 5 x 3 Sample Windows

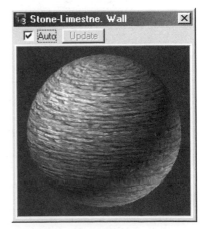

3. Double-click on any sample window, and a larger version of it will appear in a floating window. You can move and resize this window. It is for closer examination of the results of changing settings in that material's parameters

4. Close the floating sample window

Material: Floor and Landing

The last items that need materials are the Floor downstairs, and the Landing upstairs. Both would look good with aged wood planks.

1. Type H, select the Floor and the Landing

2. Select Get Material

3. In the Material / Map Browser, verify that the Browse From field is still set to Mtl Library, then in the File field, choose Open, and open Wood.mat

4. Drag material Wood-Old Planks onto an unused sample sphere

5. Choose Assign Material to Selection

6. Close the Material Editor and the Material / Map Browser

7. Save the scene

Mapping Coordinates

Most of the materials you have assigned in this scene make use of bitmaps somehow, whether it be to show a pattern on a surface, enhance the look of bumpiness, or control how the surface shines. The renderer needs to know how each bitmap is to be applied to each surface. The instructions for bitmap application to a surface are called Mapping Coordinates.

Mapping Coordinates are just a set of three dimensions oriented to an object. Mapping Coordinates can be assigned at creation for most parametric objects, and they can also be assigned to any object by applying a UVW Map modifier to the object. If Mapping Coordinates are assigned via the Modifier, you can see and manipulate a frame in the views that gives a visual representation of the image "projector" (that frame is called a Gizmo). So Mapping Coordinates are a set of instructions for applying a bitmap to an object, and they are shown as a Gizmo of a certain size and orientation on the object.

Mapping coordinates work together with Tiling to determine the scale of a bitmap on a surface. Tiling is how many times the bitmap image repeats across a given dimension. If a UVW Map Modifier has been applied to the object, then Tiling by default repeats once across each dimension of the Gizmo. The size of the Gizmo will be the size of one repetition of the bitmap. Set the Tiling to 5, and the bitmap will repeat five times across the Gizmo. So to change the scale of a pattern, you can either change the size of the Mapping Gizmo, or you can change the amount of Tiling.

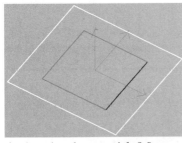

A simple plane, with Mapping Coordinates assigned (the inner, dark rectangle)

A bitmap applied to the plane, according to the Mapping Coordinates. Tiling = 1

The same bitmap, using the same Mapping Coordinates, with Tiling set to 3

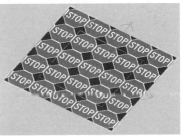

Assign Mapping Coordinates to the Roof

Most of the objects in the scene already carry Mapping Coordinates, either because they are parametric objects and you checked Generate Mapping Coordinates as you created them, or, as in the case of the floor, you used a UVW Map Modifier in modeling the object. Only the roof lacks Mapping Coordinates, as you'll see when you try to render.

1. Make the Camera view active, move the Time Slider to frame 97

2. Select Quick-Render. A dialog box appears warning that the roof is missing Mapping Coordinates

3. Cancel the rendering process

4. Select the roof, and in the Modify panel, apply a UVW Map modifier

Why the name "UVW Map"? The creators of VIZ did not want to call it "XYZ Map", for semantic reasons, so they backed up three letters in the alphabet.

5. In the UVW Map parameters, set Mapping type to Box

6. Quick-Render the Camera view

You should have a ceiling with a coarse stucco texture:

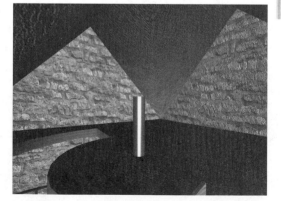

7. Make the Perspective view active, navigate to get a good view of the roof, and render the scene

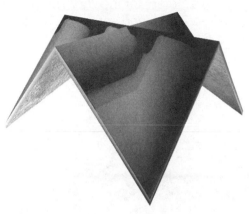

Those are extremely large spanish tiles. The Mapping Gizmo is too big.

Box Mapping Coordinates use the six planes of a box as six projectors of the bitmap, "shooting" the bitmap in from six directions onto the object. Each of the triangular faces that makes up the object picks up its mapping instructions from whichever of the six planes it is most parallel to. This makes Box Mapping an excellent choice for many architectural objects because, for better or for worse, most architecture is fairly boxy.

Bitmap of Spanish tile

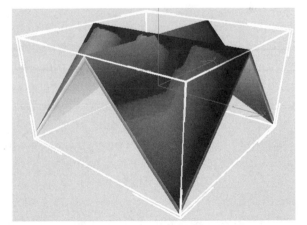

Six-sided Gizmo that Box Mapping employs to project the bitmap onto the object

Size the Mapping Gizmo

8. In the Modify panel, in the parameters for the UVW Map Modifier, set the Length, Width, and Height spinners to 2 feet, and set the U Tile spinner to 0.5 (no units), and the V Tile spinner to 1.5. Re-render. The scale of the tile should look better:

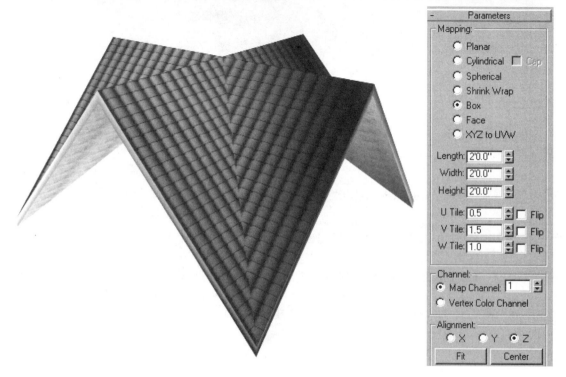

These numbers for the Mapping dimensions and the tiling were the result of trial and error. You could also set the numbers to exact real-world values. The Spanish tile bitmap shows six tiles– two rows of three tiles. Maybe each course of tile is 10 inches, and each tile is 6 inches wide, so the picture represents 18 inches across and 20 inches vertical. The right combination of 18 inch and 20 inch values in the Length, Width, and Height spinners of the Modifier should produce properly scaled tiles. Once you've established the proper scale for a bitmap, there's a Modifier called Map Scaler that locks that scale, allowing you to resize the geometry without ruining the scale of the bitmap.

Be aware that in many cases the proper scale will be too small to look good in the rendering. Don't be afraid to make slightly oversized bricks, if it makes a better picture. The ability of your eyes and your mind to discern and conjure all the detail of brickwork on a distant building cannot be approached in a rendering. You're better off achieving a compelling suggestion of brick than a confusing overdetailing.

146

Lights: Target Direct and Omni

To finish this chapter, place a couple Lights in the scene: one for general illumination of the interior of the tower, and another to throw shadows through the doors and windows.

1. Make the Top view active, then Zoom and Pan so that there is plenty of room to the southeast to place a light source

By default, the Zoom tool is set up to zoom about the point where you first begin to drag. This allows you to pan and zoom at the same time. In this example, begin the Zoom by clicking near the upper-left corner of the tower. That corner of the tower will remain near the upper-left of the viewport, and the Zoom will show space to the lower-right (Southeast) of the tower

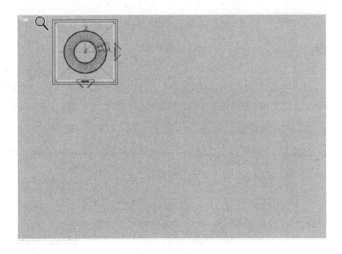

Target Direct Light

A Direct Light is either a disk or a rectangle that casts parallel rays of light, as opposed to a Spot Light, which casts a cone of light. Direct Lights are appropriate for the Sun. Target Direct Lights have both a Light object and a Target object, so they are easy to place properly.

2. In the Create panel, choose the Lights category, then choose Target Direct

3. Drag a Target Direct Light from a point southeast of the tower, and release over the center of the tower

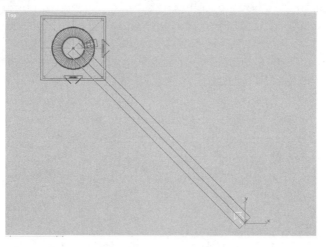

4. Make the Front view active, Zoom Extents, then move the light up off the ground plane. Keep the angle fairly low to the ground. Envision the shadowing effect you're trying to create – long, dramatic shadows from the door and window openings and mullions. The lower the light's angle, the longer the shadows

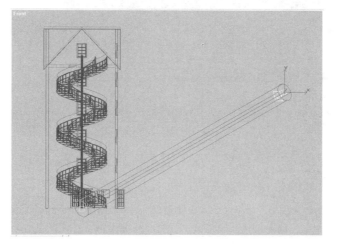

5. Switch to the Modify panel

6. Rename the light $Direct SE

A dollar sign at the front of the names of lights will cause the lights to appear together in the list, near the top of the list, when you use the Select Objects dialog box. For the same reason, put a # sign at the front of camera names. To be able to quickly select all objects outside of a building, you might add the underscore at the beginning of the name of any outdoor object (_shrub01, _tree01, etc). Items beginning with an underscore will appear in the list just below items beginning with a symbol. Use whatever tricks work best for you when naming things, but do try to develop a scheme, or you'll waste much time scrolling through the Select Objects list in search of object names.

1. Set the Light's Hue spinner to 40, the Saturation spinner to 20, and the Value spinner to 255

2. Set the Multiplier spinner to 1.25

When the Value is all the way up, and the light is still not bright enough, you can boost the Value with the Multiplier.

3. Open the Directional Parameters rollout

4. Set the Hotspot spinner to 50 (the Falloff spinner will adjust automatically to stay slightly larger than the Hotspot)

The Hotspot is the brightest part of the pool of light. The intensity of the light fades at the edges, until there is no illumination at the Falloff distance.

5. Open the Shadow Parameters rollout

6. Set the shadow type drop-down to Ray Traced Shadows

This light will be traveling through a material with transparency– the glass panes in the doors and windows. Ray traced shadows are the only ones that handle transparency correctly.

"Light's- Eye" View

1. Make the Front view active, and type the dollar sign (shift-4). You will now be looking through the Direct Light, seeing what it sees

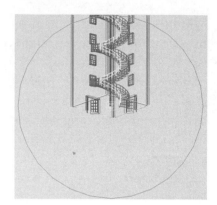

You can make a "light's-eye" view for any Direct or Spot Light. With the light view active, notice that the view navigation tools have changed, to tools pertinent to lights.

2. Use the Truck Light tool in the view navigation tools to center the tower in the light view

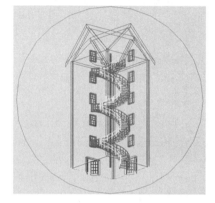

3. Once the tower is centered in the light, type F to change the light view back to a Front view

4. Make the Camera view active, drag the Time Slider to frame 0, and render

Your scene should look something like this:

Notice the position of the light from the door- it meets the wall and travels partway up. If you do not see the same effect in your scene, move the light around in the Top and Front views and re-render the scene until the light is positioned correctly.

Omni Light

1. Make the Top view active, then use Region Zoom to zoom tight to the tower

2. In the Create panel, Lights category, choose Omni

3. Position the cursor in the Top view over the Stairs, a bit south of east, and click to place the Omni Light, as shown here:

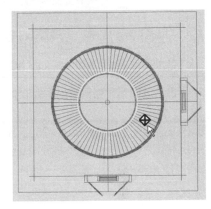

4. Make the Front view active, choose the Move tool, then right-click on the Move tool

5. In the Transform Type-in, enter 13′ in the Offset: Screen Y: spinner, hit Enter. The light should move 13′ off the floor. Close the Transform Type-In

6. In the Modify panel, set the light's parameters:

7. Rename it $Omni-Interior01

8. In the HSV spinners, set H = 0, S = 0, V = 120

9. Set the Multiplier to 1

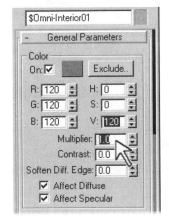

Shadow Map Parameters

10. Open the Shadow Parameters rollout

11. Open the drop-down list and switch to Shadow Map

12. Open the Shadow Map Params rollout

13. Set the Size spinner to 400

14. Set the Smp Range spinner to 8

Remember that Shadow Map Size is the pixel dimensions of the bitmap used to depict the shadows, if you could open that bitmap in Photoshop (in this case, the bitmap would be 400 x 400 pixels).

Sample Range is the distance at the edge of the shadow across which the shadow fades. Low Sample Range distances make crisper shadows, and high Sample Ranges make softly defined shadows. For a Shadow Map, 400 is a low value, and 8 is a fairly high Sample Range. The shadows that these settings will produce will be soft-edged.

Good Shadow Mapping involves finding a balance between Map Size and Sample Range. When adjusting Map Size and Sample Range, you look for the combination of values that gives the crispness or softness you want, while not allowing the edges of the shadows to get grainy. The combination of low Map Size and high Sample Range often causes the shadows to get rough and grainy at the edges. It takes trial and error to get shadows very soft-edged, but still sufficiently defined. You can add great realism and life to an exterior view of a building facade by shining a shadow-casting light across the facade at a fairly small angle (more parallel to the wall than perpendicular), and setting that light to cast Shadow Maps with very low size (a few hundred) and very high Sample Range (maybe 12 to 15). The result is very subtle shadowing in the areas of relief on the facade, which adds depth to the facade, softens it a bit, and adds a suggestion of age to the building.

Map Bias can be thought of as a rubber-band connecting the object and its shadow. If Map Bias is too high, the rubber-band is too long, and the shadow will separate from the object. The shadow of a chair leg will start a couple inches beyond the leg, so that it looks as if the chair is floating above the floor. If the Map Bias is too low, the rubber-band is so tight that the shadow is actually drawn in front of the object– you will see the shadow of the chair leg starting closer to the light source than the leg is. This is not a setting you need to change often, and it usually needs to be lowered, to as low as .1.

14. Render the Camera view. The scene should look something like this:

Now you need some illumination in the top of the tower.

15. With the Omni Light still selected, choose Edit / Clone from the menus. In the Clone Options dialog box, choose Copy as the clone type, then choose OK

16. Make the Front view active

17. Choose the Move tool, then right-click over it

18. In the Transform Type-In, enter:
       Offset: Screen X: -2
       Offset: Screen Y: 54
       Offset: Screen Z: -8
       (note the minus signs)

19. Close the Transform Type-In

20. Drag the Time Slider to frame 96

21. Render the Camera view

22. Save the scene

Render a Movie

This will take some time, so set up for this part and render the animation over your lunch hour or overnight, as you will likely not be able to use your computer during the rendering process. While it is possible to use other programs while rendering, you should not count on it, as your computer's performance may degrade intolerably.

1. With the Camera view active, select the leftmost of the three Render teapots to open the Render Design dialog box

2. At the top of the dialog box, set the Time Output choice to Active Time Segment

3. Set the Output Size to Width = 400, Height = 300

The larger the pixel dimensions of the movie, the slower it is likely to play back.

4. In the Render Output field, select the Files button

5. In the Render Output File dialog box, browse to the C:\Viztutorials\Chapter3 folder as the Save In location

6. Enter Tower as the File Name

7. Open the Save As Type dropdown list and choose AVI File

8. Select Save, and the Video Compression dialog box will open. Leave the compression settings at the defaults and select OK

9. Open the VIZ Default Scanline A-Buffer rollout, verify the Mapping, Shadows, Anti-Aliasing, and Filter Maps are all checked

10. Select the Render button

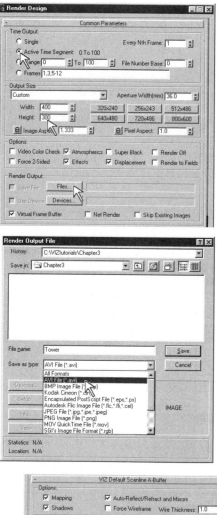

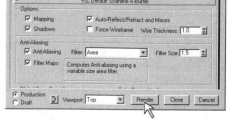

On a computer with a Pentium2-266 processor and 256 megs RAM, the animation requires just over an hour to render. When finished, the animation will be opened for playback. For playback later, use Windows Media Player, or in VIZ, choose Tools / Display Image.

The animation will play extremely quickly, because it is only 100 frames in length. To ascend the stairs at a more reasonable rate, the animation length would need to be 300 frames, so that playing back at 15 frames / second, the ascent would take 20 seconds.

Summary

This is a good start on the interior lighting scheme. There is still much to do to light the tower properly, but further exploration of lighting can wait until a later chapter.

This model of the tower is very simple: walls, floors, a roof, some doors and windows, and a spiral stair. The lighting system is simple too. But the rendering is compelling—there is a mood conveyed, and you can imagine this tower as the setting for a story. There are two things that are vital in making this happen. The first is that the interior of the tower is dark. The light through the doors and windows creates high contrast in the dark space. Highlights on the spiral stair also read more dramatically in the dark than they would in a bright room. So contrast is important in an image. The second thing that makes the scene interesting is that the spiral stairs were chosen specifically for the curved parts. It is very easy to make dramatic highlighting happen on curved surfaces, and it can be quite difficult on flat surfaces. Really lively, realistic flat surfaces are probably the most difficult thing to accomplish in VIZ.

So if you want to make a simple model and have people be excited by it, keep these two things in mind: darker spaces with high contrast are more evocative, and curved objects in the room are an easy way to add nice highlights. Look for these two factors in images shown in magazines, contests, and ads for 3D software. Look at the image on the box in which VIZ 3 is shipped, and flip through the Kinetix Resource Guide brochure that ships with VIZ. There are lots of dark scenes and curvy, shiny things. When you see an image of a brightly-lit scene comprised mostly of flat surfaces, and that scene looks really alive and is a place you want to be, then be impressed. It takes a lot of work to get a scene to that state.

Furnishings: Shapes, Modifiers, Lofting

In this chapter you'll build furniture using a variety of techniques, including:

- Mesh Edit a Standard Primitive
- Bevel a Shape
- Bevel Profile a Shape
- Lathe a Cross-Section
- Loft a Shape

These tools are not specific to furniture, and by the end of this exercise you will have the skills needed to model almost anything you are likely to encounter in a building. The secret to modeling buildings quickly and realistically is not so much to learn a dizzying array of tools and techniques as it is to learn good habits for efficiency and clever ways to maximize the versatility of a few powerful tools. One of the most satisfying aspects of modeling with VIZ is that if you use your imagination and avoid rigid thinking, you'll discover uses for tools that may not have occurred to anyone else. Creativity is the competitive edge in VIZ.

This is a lengthy chapter, but because it involves the making of distinct units of furniture, it is really a series of modules, with several good stopping points in the exercise.

Primitive: Box

1. Start a new scene (File / New)

2. From the menus, choose Tools / Drafting Settings / Units Setup

3. Set Units to US Standard, Feet w/Decimal Inches
   Set Default Units = Inches, choose OK

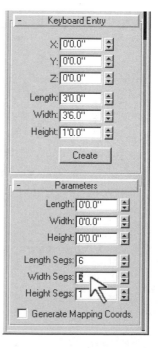

4. Make the Top view active

5. In the Create panel, Geometry category, choose Box

6. Open the Keyboard Entry rollout and set:

> Length = 36
> Width = 42
> Height = 12

7. In the Parameters rollout, set:

> Length Segs = 6
> Width Segs = 6
> Height Segs = 1

8. Choose Create

9. Zoom Extents All

10. Turn the Grid off in all views

11. Use the F3 key to set all views to wireframe display

12. Make the Perspective view active and Arc-Rotate to get a good view of the Box

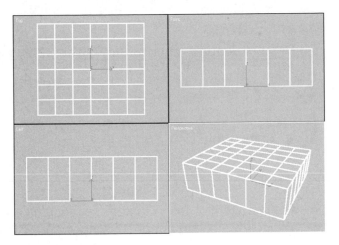

Right-click Menu

One of the biggest changes in the VIZ 3 interface is the expansion of the pop-up menu that appears when you right-click in a view. In VIZ 2, if you right-clicked over an object, a pop-up menu appeared with a half-dozen functions. In VIZ 3, this menu now contains a couple dozen functions. There are controls for Transforms, there are Modifiers that can be added right off the menu, and perhaps most useful, there are numerous mesh editing tools available in the menu. This chapter uses mesh editing extensively, and rather than constantly scrolling up and down the Modify panel to do things like change Sub-Object mode, you will use the right-click menu. You'll find that mesh editing is more fun and interactive this way.

The right-click menu does not always appear at the first right-click. If you are in a view navigation mode or an object creation mode, the first right-click will cause you to leave that mode, and the second will open the menu.

1. Right-click over the Box, and from the pop-up menu, choose Convert to Editable Mesh

Mesh Editing

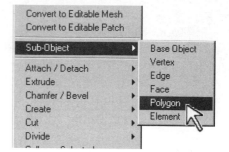

2. Right-click again, choose Sub-Object / Polygon

3. Make the Front view active

4. At the bottom of the VIZ interface, set the Crossing / Window Selection button to Window Selection

⚠️ Even if the button is already set to Window Selection, click it twice to reset it to Window Selection– there seems to be a problem with this button. If selections are not working correctly, try resetting it.

5. In the Front view, drag a selection window to select only the polygons on the top of the Box

6. Type W to maximize the Front view

7. Right-click over the Box, choose Chamfer / Bevel, then Bevel Polygon

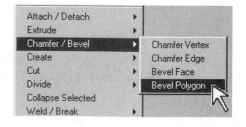

8. In the Front view, position the cursor over the selected polygons, so that it shows a Bevel cursor. Drag upwards to set a small height (about half an inch), release, then move the cursor upwards to angle the new faces outward at about a 45 degree angle, and click to set the angle

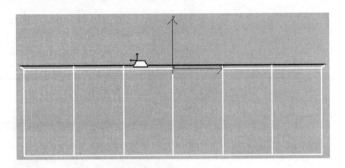

9. Drag another small bevel out, this time angling the bevel inward, back to the original dimensions of the box

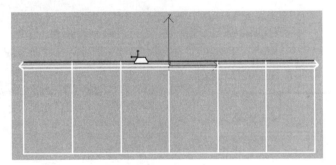

The faces you have just made will be a wooden bead that goes around the seat at the top of the seat's base.

10. Type W to return to four views, then make the Perspective view active and maximize it

11. Right-click over the box, and from the pop-up menu, choose Select Mode

12. Hold the Alt key on the keyboard, and deselect faces so that only the faces where the arms and back attach are selected, as shown here:

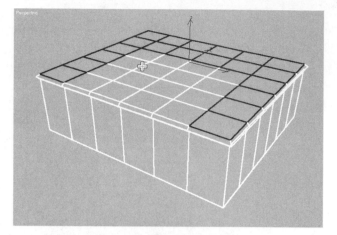

To deselect faces, pick in the middle of each quadrilateral (square).

13. Right-click, and from the pop-up menu, choose Extrude / Polygon

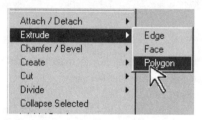

14. Place the cursor over any of the selected faces so that it shows the Extrude cursor, then drag upward to extrude new faces that are roughly cubic, as shown here:

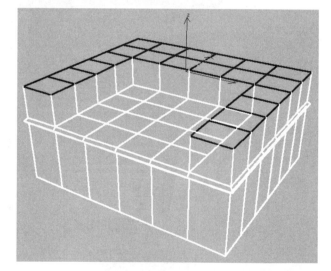

15. Right-click, and from the pop-up menu choose Chamfer / Bevel, then Bevel Polygon

16. Drag upward to extrude a new set of faces roughly half the height of the previous Extrude, release the mouse, then move the mouse upward to flare the sides of the new faces out about 30 to 40 degrees, as shown here, then click to set the angle

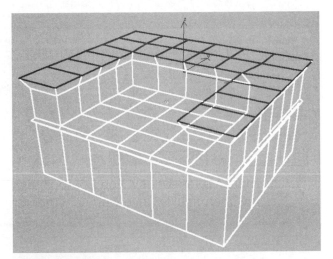

17. Bevel another set of faces, roughly the same height as the last set, and angle these inward about 50 to 55 degrees, as shown here:

18. Right-click, and switch to Sub-Object Vertex

Sub-Object	▶	Base Object
		Vertex
Attach / Detach	▶	Edge
Extrude	▶	Face
Chamfer / Bevel	▶	Polygon
Create	▶	Element
Cut	▶	

19. Return to four views, make the Front view active

20. Zoom Extents All

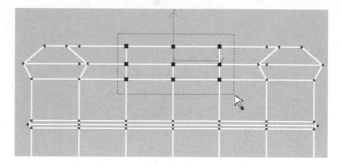

21. Drag a selection of vertices at the back of the seat, as shown here:

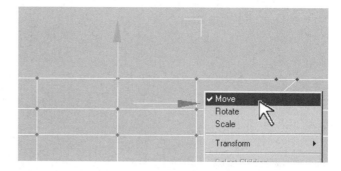

22. Right-click, and choose Move from the pop-up menu

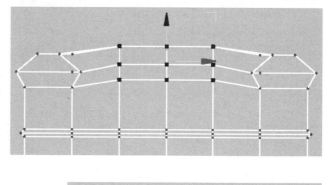

23. In the Front view, position the cursor over the Y axis of the Transform Gizmo, then move the selected vertices up a bit, as shown here:

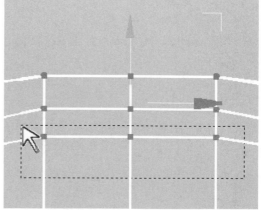

24. Hold the Alt key on the keyboard, and drag a selection window to deselect the bottom row of selected vertices

25. Move the two selected rows of vertices up a bit more

26. Hold the Alt key, and deselect the bottom row of selected vertices, leaving only one row selected

27. Move those vertices up a bit

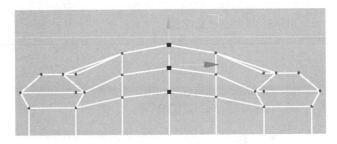

28. Finally, select the center column of vertices and move them up some, as shown here:

29. Make the Perspective view active

30. Right-click, and choose Sub-Object Polygon

31. Right-click, and choose Select Mode

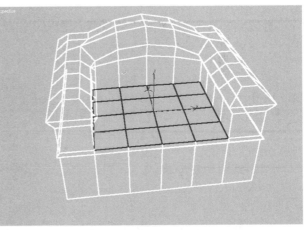

32. Select one of the Polygons of the seat cushion area, then hold the Ctrl key and select the rest of those Polygons. You will need to Arc-Rotate to be able to select all the Polygons

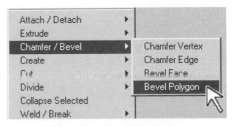

33. Right-click, and choose Chamfer / Bevel, then Bevel Polygon from the pop-up menu

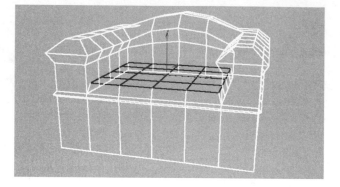

34. Drag to extrude the seat cushion up a few inches, release, then set a small angle to create a gap at the sides of the cushion

Material ID Numbers

The last thing to do before smoothing the seat into its final shape is to assign the faces that will be the wooden trim bead one Material ID number, and the faces that will be leather another Material ID number.

35. Make the Front view active and maximize it

36. Right-click, and choose Select Mode from the pop-up menu

37. At the bottom of the VIZ interface, set the Crossing / Window Selection button to Window Selection

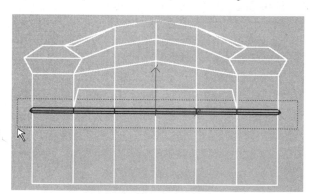

38. In the Front view, drag a narrow selection window to select those faces that make up the wooden bead. If you need to, Zoom in to fine-tune your selection. Use the Alt key to deselect if you select too many faces

39. Scroll to the bottom of the Modify panel, and set the Material ID spinner to 2

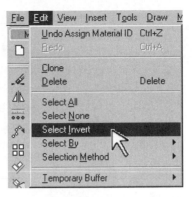

40. From the menus at the top of the screen, choose Edit / Select Invert, to deselect the faces of the wooden bead, and select all the other faces of the seat

41. Set the Material ID number for the rest of the seat to 1

42. Right-click, and from the pop-up menu, select Sub-Object / Base Object

43. Type W to return to four views

44. Save the scene as Furnishings01, in the C:\Viztutorials\Chapter4 folder

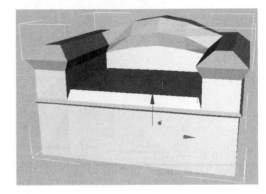

45. Make the Perspective view active, then type the F3 key to set the view to Smooth + Highlights

Meshsmooth Modifier

The Meshsmooth Modifier eases and rounds the edges of the mesh. To do so, it has to add faces, so you compromise between better-looking objects and slower rendering times and display speed. In VIZ 3, powerful new controls have been added to Meshsmooth. The name given to the new controls is NURMS (Non-Uniform Rational MeshSmooth), which seems like an overly technical name for the simple idea of weighted vertices. It simply means that Meshsmooth is applied via a lattice surrounding the object, and each vertex of that lattice is assigned a weight, or pulling force, that you can alter to accurately reshape the object. This feature makes Meshsmooth one of the most powerful modifiers, allowing you to sculpt an object in a very interactive, but still parametric, way.

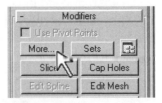

1. In the Modify panel, apply a Meshsmooth modifier (you will need to choose the button labeled More, found above the default 10 modifiers, to find Meshsmooth)

2. In the Modify panel, find the spinner labeled Iterations, and set it to 2, to make the seat smoother

3. In the Modify panel, in the group labeled Surface Parameters, check the box labeled Separate by: Materials

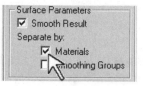

The Separate by: Materials option prevents Meshsmooth from softening the seams where different materials meet. You assigned Material ID #2 to the faces that will be the wooden bead. You want to keep the boundaries of that bead crisp and unaffected by Meshsmooth. That is what Separate by: Materials does.

The seat looks fairly good, but the fronts of the arms are a bit too bulgy. Use the new vertex weighting feature of Meshsmooth to size them down some.

NURMS- Vertex Weighting

1. In the Modify panel, turn on Sub-Object

2. Make the Front view active

3. With the Ctrl key held down, drag two selection windows to select the vertices of the Meshsmooth lattice at the arms, as shown here:

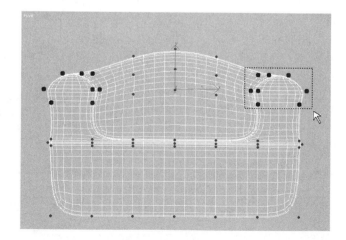

4. Make the Top view active

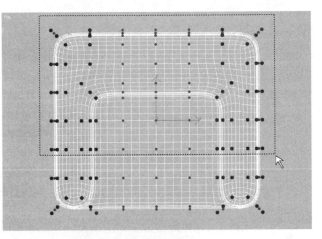

5. With the Alt key held down, drag a selection window to deselect all but the vertices at the front of the arms

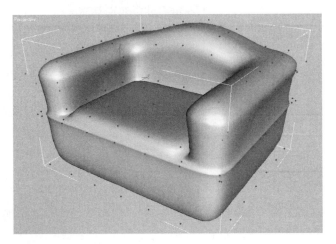

6. In the Modify panel, find the spinner labeled Weight, and enter .1

7. Turn off Sub-Object

8. Save the scene

The arms are less bulgy now:

One of the most common questions that new users of VIZ post on support forums (such as the one available at www.vizonline.com) is "where is the 3D chamfer / fillet tool?" There is no modifier called 3D Chamfer / Fillet, and there should be. The problem with using Meshsmooth as a 3D chamfer / fillet is that edges are not chamfered at a consistent forty-five degrees. The relative dimensions of the faces meeting at an edge affect the angle of the chamfer. Meshmooth will chamfer a cube perfectly, but applying it to a long thin box produces uneven chamfers, with larger faces being chamfered more, and shorter ones being chamfered less. Hopefully a 3D Chamfer / Fillet modifier will be added in a future release of VIZ.

Multi / Sub-Object Material

Finish the Easychair model by assigning a material. This will be a Multi / Sub-Object material, with one Submaterial being leather, and the other being dark wood for the trim bead.

1. Open the Material Editor

2. With the first sample sphere active, choose the material Type button labeled Standard, to open the Material / Map Browser

3. In the Browser, choose Multi / Sub-Object, choose OK

4. In the Replace Material dialog box, choose Discard Material, choose OK

5. Choose the button labeled Set Number

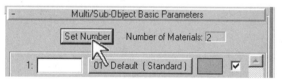

6. In the Set Number of Materials dialog box, enter 2 in the spinner, choose OK

7. Name the top level of the material Easychair

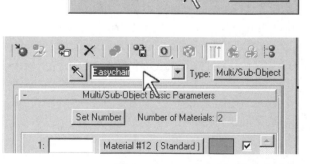

8. Choose the Get Material button

9. In the Material / Map Browser, choose Browse From: / Mtl Library

10. Choose File / Open

11. In the Open Material Library dialog box, choose Skin.mat

12. Drag material Skin-Tan Leather from the Browser to the first submaterial button

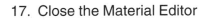

13. In the Browser, choose File / Open, and open 3dsviz.mat

14. Drag material Wood-Bubinga to the second submaterial button

15. Close the Material / Map Browser

16. Choose Assign Material to Selection

17. Close the Material Editor

18. In the Modify panel, apply a UVW Map modifier

171

19. Set the Mapping Type to Box

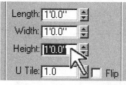

20. In the UVW Map parameters, set the Length, Width, and Height to 1'

21. Render the Perspective view

22. Save the scene

Pedestal and Vase: Bevel Modifier

Mesh editing is actually one of the more complicated ways to make furniture. Many furnishings can be made by applying simple modifiers to shapes, and one of the simplest modifiers is Bevel. To bevel is to extrude at an angle. This modifier is frequently used to make beveled text in logos, but is also a workhorse in architectural modeling.

You will use the Bevel modifier to make a wooden pedestal and a glass vase to display on the pedestal.

1. Make to Top view active

2. In the Create panel, Shapes category, choose NGon

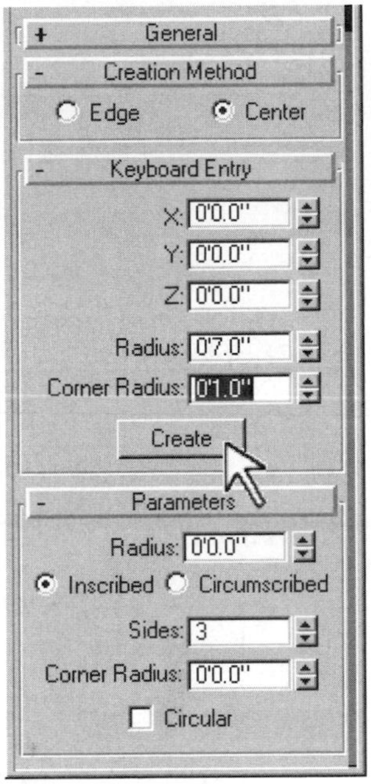

3. Open the Keyboard Entry rollout, and enter a radius of 7″ and a Corner Radius of 1″

4. In the Parameters rollout, set the Sides spinner to 3

5. Choose Create

6. Move the NGon to one side of the leather seat

7. Zoom Extents All

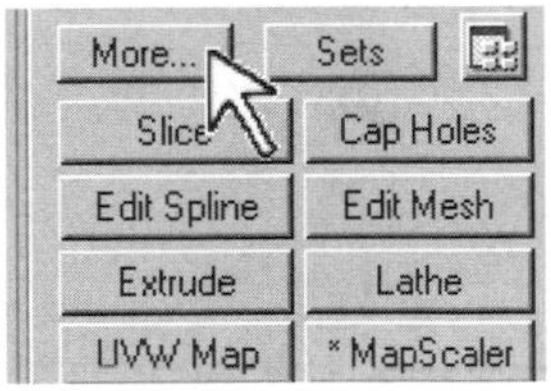

8. In the Modify panel, choose the button labeled More, above the 10 default Modifier buttons

9. In the Modifiers list, select Bevel, choose OK

When you apply the Bevel Modifier you will see the NGon become solid in any shaded views. There are many modifiers that turn a shape into a planar surface, or mesh: Extrude, Bevel, UVW Map, Normal, Edit Mesh, and Mesh Select.

10. In the Bevel parameters, uncheck Capping / Start, to remove faces on the bottom of the pedestal

11. Check Generate Mapping Coordinates

12. In the Bevel Values rollout, set:

      Level 1 Height = 1′6″
      Level 1 Outline = -1″

      Put a check in the Level 2 check box

      Level 2 Height = 1 ′
      Level 2 Outline = 1.5″

      Put a check in the Level 3 check box

      Level 3 Height = .5″
      Level 3 Outline = -.5″

As used here, "outline" means "offset".

The pedestal (which you should name Pedestal) should look like this:

The half-inch chamfer that the Level 3 settings added at the top of the pedestal is a good example of one of the best modeling habits you can develop in VIZ, which is to look for ways to add chamfers or fillets to things, to create edges that will pick up the light and help add definition and visual interest to the object. Anytime you are about to apply an Extrude modifier to a shape to make it 3-dimensional, stop and consider whether Bevel might not be a better choice, as it will easily allow you to add a chamfer or fillet to the object.

Now for a glass vase to display on the pedestal. You will apply a Bevel modifier to a Circle to begin modeling the vase, and you'll see that Bevel works as well for curved objects as it does for straight-sided ones.

1. Make the Perspective view active, and Arc-Rotate, Pan, and Zoom to get a good view of the top of the Pedestal. Type the F3 key to set the Perspective view to a wireframe display mode

2. At the Status Line, choose AutoGrid

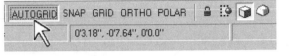

3. In the Create panel, choose Circle

4. Drag a Circle on the top of the Pedestal

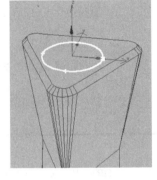

AutoGrid creates a temporary construction grid parallel to whatever surface you click on. With AutoGrid, you will not need to do as much repositioning of objects after you have created them.

5. Turn off AutoGrid

6. Switch to the Modify panel and set the radius of the Circle to 2.5″

7. Apply a Bevel modifier to the Circle, then Zoom Extents Selected

The vase is quite tall at this point. The first time you apply a Bevel modifier in a scene, the default values will all be zero. When you apply another Bevel modifier to something else, the parameters you set for the previous Bevel will be carried over as the defaults for the next Bevel.

7. Check Capping / Start, and Uncheck Capping / End

These Capping settings will make the vase closed at the bottom and open at the top.

8. Change from Linear Sides to Curved Sides

9. Set Segments to 4

10. Check Smooth Across Levels

11. Check Generate Mapping Coordinates

12. In the Bevel Values rollout, set:

> Level 1 Height = 6″
> Level 1 Outline = 3″
>
> Level 2 Height = 2″
> Level 2 Outline = -1″
>
> Level 3 Height = 1"
> Level 3 Outline = -2″

13. Name the object Vase, and save the scene

Your vase should look like this:
If there appear to be faces missing at the top, there aren't- those faces face away from your point of view (backfaces), so you do not see them.

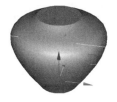

Now be creative; experiment with the various Bevel Height and Outline values to re-shape the vase any way you want. You will find that the three levels do not behave exactly as you would expect from the values that you enter in the spinners. The three levels do not act independently; they affect each other. If you begin by setting Level 1 Height to 3″, then set Level 2 Height to a foot, you will see in the object that Level 1 does not seem to be 3″ tall. It gets squashed down, and may actually fold back over itself. So, when using Bevel to make vases and dishes, just play with the values by eye until something good happens.

Bevel is a very versatile modifier. It is perfect for tall buildings with parapet roofs. Level 1 Height is the height of the building. Level 2 Height is zero, and Level 2 Outline is the thickness of the parapet wall (a negative value). Level 3 Height is a negative value, the height of the parapet wall. Level 3 Outline is zero.

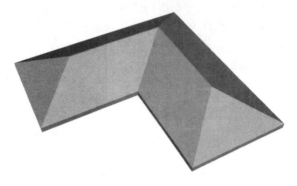

Bevel is also good for hip roofs or mansard roofs. Level 1 Height is the fascia dimension, Level 2 Height is the distance from top of fascia to peak of roof, and then set Level 2 Outline interactively (to some negative value) until the roof comes to a peak.
In the example shown here, both sides of the L plan are the same dimensions. When they are not, you will need to apply an Edit Mesh modifier after the Bevel modifier, choose Sub-Object Vertex, select pairs of vertices that do not quite meet at the peak, and choose Collapse in the Edit Mesh parameters to join vertices into a perfect peak.

Oak Table: Bevel Profile Modifier

Once you've found a few uses for Bevel, you will find yourself wishing that the Bevel modifier had more than three levels, because with six or eight you could model a huge variety of objects. That's where Bevel Profile comes in. It is like Bevel, except that instead of defining Levels and Outlines, you draw a 2D line shape that defines all the heights and offsets (the silhouette, basically) of the object.

The heavy wooden table you will make using Bevel Profile is accomplished by drawing the oval footprint of the table in the Top view, then drawing the complex profile in a side view, then assigning the Bevel Profile modifier to the oval, and choosing the complex Line shape as the profile.

1. In the Top view, beside the leather seat, drag a Rectangle, of no particular size

2. Name the Rectangle Table

3. In the Modifiy panel, set the parameters of the Rectangle:
   Length = 2′
   Width = 3′6″
   Corner Radius = 1′

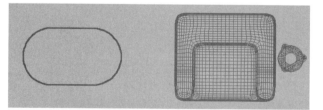

4. In the Front view, Pan and Zoom to see the radiused rectangle on edge, and some space above it. Draw a Rectangle of any size, then adjust its parameters to Length = 3′8″, Width = 2′. Position the rectangle anywhere above the table base shape, and zoom to see it

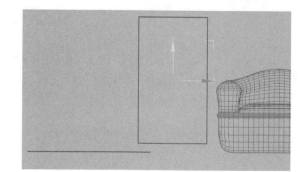

Drawing with the Line Tool

So far you have drawn only simple shapes like Circle and NGon. To draw the profile of the table you will use the Line shape, which is a bit more complicated. If you've used a vector illustration tool like Adobe Illustrator, the Line shape will be fairly familiar. If you are new to drawing with Bezier lines, it will take a bit of getting used to.

1. Maximize the Front view

2. In the Create panel, choose Line

3. Near the bottom of the Create panel, set Drag Type to Smooth

"Initial Type" and "Drag Type" are two types of mouse clicks. Initial Type is a quick click with no dragging action, and Drag Type means that you drag the mouse a bit as you click. If your Line shape is to consist entirely of straight segments, then you will only need the Initial Type. If your Line shape is to have some straight and some curved segments, you will be using both types of clicking actions (as is the case for this profile).

4. Using the rectangle as a guide, draw a shape similar to the one shown below. The shape has ten vertices. In the diagram, Initial type vertices are labeled with an "i", and Drag type with a "d". As you draw, keep these things in mind:

The exact shape does not matter at this point. Just get the general shape, using ten vertices.

Use the Backspace key on the keyboard to "undraw" mistakes. You can back up as many segments as you want.

A right-click will finish the shape. If you accidently right-click too early, just delete the partial line and start again.

Always draw in a counterclockwise direction. In this case this means start at the bottom and work your way up.

Align the vertices of the straight segments as well as you can by eye.

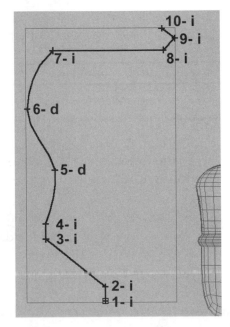

179

If drawing the line does not go as smoothly as you might hope, relax. Adding the curved segments takes some practice, and it's tough at first to sketch in such a disconnected way, with your eyes up on the screen and your hand moving unseen on the desk. Just remember that you don't need to get the shape perfectly correct as you draw– anything close will do. You have total control over every part of the line later in the Modify panel.

Spline Editing

Once the basic shape is roughed out, you switch to the Modify panel to fine-tune the shape. You can control the location of any vertex in the shape, and you can control the curvature (tangency) of every segment as it approaches and as it leaves a vertex.

1. With the Line complete, switch to the Modify panel

2. Name the Line %Table Profile

The % symbol is a selection strategy. By the end of the chapter there will be several 2D shapes in this scene that are not part of the rendered image, but are used in modeling various 3D objects. If each of those shapes has a symbol at the start of its name (any symbol will work), the shapes will be grouped together at the top of the Select by Name dialog box, for quick and easy selection.

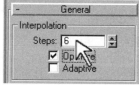

3. Open the General rollout and set the Steps spinner to 6

Steps will be explained in a few pages.

4. In the Selection rollout, choose the red shortcut button for Vertex, to drop into Sub-Object Vertex mode

As you choose Vertex, notice that the Sub-Object button above turns yellow, and that the drop-down list next to the Sub-Object button reads Vertex. Remember that in making the leather easy chair you had to switch frequently between Sub-Object Polygon, Sub-Object Vertex, and Sub-Object/Base Object. In that exercise you changed Sub-Object modes via a right-click menu. It doesn't matter which method you use to navigate Sub-Object modes. You can right-click in the view to access the menu, you can choose the yellow button and the drop-down list, or you can use the red shortcut buttons. Do whatever seems most efficient to you.

5. Drag a window around the entire shape to select all of the vertices. At each vertex you will see a Transform Gizmo

6. The Transform Gizmos are not needed in editing this spline, and they are a distraction. The best way to turn off Transform Gizmos is to just type X. The other way is to choose, from the menus at the top of the interface, View / Display, and uncheck Transform Gizmo

7. With all vertices selected, position the cursor directly over over any one of them and right-click to bring up a menu. From the menu choose Bezier Corner

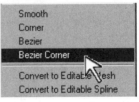

Handles with green grips appear at each vertex. With the Move tool active you can now move red vertices to alter the shape, or you can move green grips to alter how the shape behaves as it approaches and leaves each vertex. The position of each handle matters, and so does the length of the handle. Longer handles produce broader arcs, and shorter ones produce tighter arcs.

⚠ It seldom happens, but occasionally the vertices and tangency handles become locked in position, for no apparent reason. If this happens, save the scene, close VIZ, restart VIZ, and reopen the scene.

8. Click away from the shape to deselect all vertices

9. Use Region Zoom to zoom to the bottom of the shape, so you can see vertices 1 through 4

10. Select the very bottom vertex, vertex 1

Notice that this vertex has a unique appearance: it has a box around it. This designates the First Vertex, which will become important later, when you learn to create complex objects.

11. Choose the Move tool, if it is not already active

12. If the first vertex is not directly under the second, position the cursor directly over the red vertex marker and drag to move the vertex. Correct its position by eye

13. Drag the green grip on the tangent handle to a point to the right of the vertex, as shown here:

Don't worry about getting the handle exactly horizontal-if the handle looks straight (doesn't show "jaggies"), close enough.

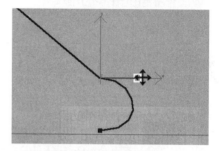

14. Select the second vertex and move its lower tangent handle to the right as well. You should now have a roughly semicircular arc at the foot of the shape. Adjust both tangent handles and vertex locations until the shape is right

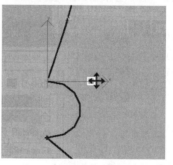

15. Move the tangent handles of the third and fourth vertices to add a bead where the foot of the table meets the leg:

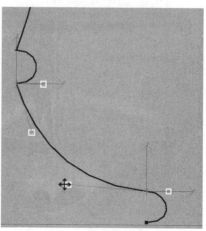

16. Adjust the tangent handles of the second and third vertices to get this curve between them:

17. Pan up to see the last three vertices of the shape

18. Work the last three vertices of the shape into a profile of the edge of the tabletop, something like what is shown here (feel free to improve on the shape if you wish)

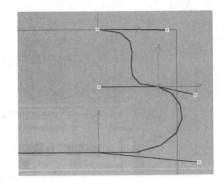

19. Move vertices and adjust tangent handles as needed to get the shape close to this appearance:

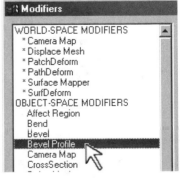

20. When you are happy with the profile of the table, turn off Sub-Object

21. Select the rectangle guide and type the Delete key on the keyboard to delete it

With the Line properly edited, you are finally ready to Bevel Profile the table.

22. Type W to return to four views, and Zoom Extents All

23. Select the Table shape (the oval footprint)

24. In the Modify panel, choose the button above the default modifiers labeled More, and from the list of modifiers choose Bevel Profile

25. In the parameters for Bevel Profile, choose the button labeled Pick Profile

26. In any viewport, click on the Table Profile

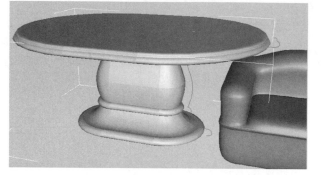

The result should be a sturdy table.

27. Make the Perspective view active and Arc-Rotate to view the table from the side. Zoom Extents Selected

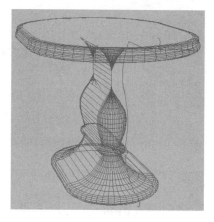

Depending on how you drew the profile shape, you may find that the table gets very skinny in places. You can edit the profile shape to change the shape of the table, and you can also change the position of the profile Sub-Object that is embedded within the table. Move the profile Sub-Object to alter the width of the table.

28. In the Modify panel, choose the Sub-Object button. The profile embedded within the table will turn yellow

29. Type X to show the Transform Gizmo (if for some reason it does not appear, from the menus choose View / Display / Transform Gizmo)

30. Position the cursor over the X axis (the red axis) of the Transform Gizmo and slowly drag along the X axis. The dimensions of the table will change. Move the profile Sub-Object until the table base has the dimensions you want

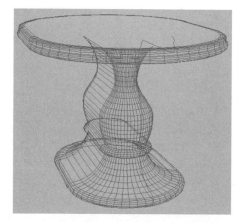

31. Turn off Sub-Object

32. Save the scene

33. In the Modify panel, uncheck Capping / Start. This will remove the faces on the underside of the Table, which will never be seen

34. Make the Perspective view active and navigate to get this view of the table top:

35. Type the F3 key to set the view to shaded display mode

In this view from above, you can see that the arcs of the tabletop aren't too smooth– they are clearly segmented. This is easily fixed.

36. In the Modify panel, open the drop-down list that currently reads Bevel Profile, and switch to the Rectangle level

This list is called the Modifier Stack. It is an editable history of the object: its state when you created it, and all Modifiers you have added to it. You can move up and down the Modifier Stack and change values in the various Modifiers, or even the creation parameters (with certain limitations), and the object will update in the views.

Shape Steps

37. Open the rollout labeled General

38. Set the spinner labeled Steps to 12

The tabletop should now have a smoother shape.

Steps is a setting that determines how many interpolations will be used between vertices of a shape when that shape is used in making an object. The default is usually six, so that an arc shape defined by a beginning vertex and an end vertex will become a mesh object defined by six facets.

39. Open the Modifier Stack drop-down list again, and return to the top of the stack, Bevel Profile

40. Arc-Rotate Selected to get a good view of the side of the table

41. Type H and Select the Line called %Table Profile

42. Type the F3 key to set the view to Wireframe

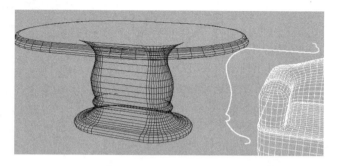

In the same way that the shape steps for the rectangle shape can be used to control the definition of arcs in the plan view of the table, the number of steps in the profile shape controls the definition of arcs in the elevation view of the table. The table has more faces than it needs to define the arcs in profile. Lower the steps to make a more efficient mesh.

43. In the Modify panel, open the General rollout

44. Set the Steps spinner to 3

45. Type the Esc key to leave the spinner, then type the F3 key a few times, to compare the table in wireframe and shaded view modes. Leave the view set to wireframe

The table is now made up of considerably fewer faces, because you lowered the interpolation between vertices of the table profile to 3, but you can see in the shaded view that the table still looks good– the curves are still reasonably smooth.

The proper number of steps for the various shapes that are used in modeling an object will vary according to the complexity of the shapes and according to how close the camera will get to the object. Keeping shape steps to a minimum is crucial to keeping the polygon count of your scene manageable.

46. Save the scene

Candlestick, Picture Frame: Lathe Modifier

Lathe is another modifier that gets applied to a 2D shape. Lathe spins the shape about an axis to produce a 3D mesh. Some typical examples of lathed objects in architectural modeling are columns, balusters, and chair legs. In these items the axis of lathing runs through the centroid of the object. In other instances you move the axis of lathing away from the centroid to produce donut-type objects, such as a rim for a bicycle wheel or the coping for a circular fountain

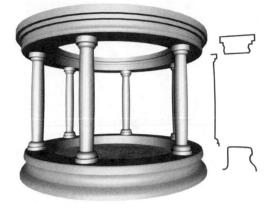

Create a candlestick using the Lathe modifier:

1. Make the Front view active, and maximize it

2. Pan and Zoom to see the tabletop and a couple feet above it

3. Draw a Rectangle above the Table, then set the parameters of the Rectangle to:

   > Length = 24″
   > Width = 2″

4. Move the Rectangle down to the tabletop, then Zoom Extents Selected

Draw the Shape to Be Lathed

5. Using the Line tool, sketch the right-hand profile of a candlestick, using the rectangle as a guide. Follow this diagram for Initial Type and Drag Type mouse clicks:

Keep the following seven things in mind as you draw:

At the bottom of the Create panel, set Drag Type to Smooth.

Don't worry about the exact shape yet, just draw the correct number of vertices in roughly the correct position.

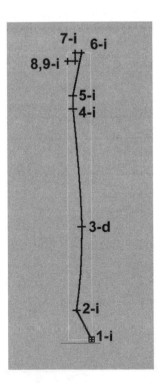

187

Draw counterclockwise, which means you will start at the bottom of this shape and work upward.

Use the Backspace key to undo segments if you make mistakes.

Finish the Line with a right-click.

Do not add any more vertices than are necessary to define the shape; excess vertices mean excess polygons later.

Use Ortho, described below, to draw any straight sections.

Ortho

The top of the candlestick is a good opportunity to practice using Ortho. Ortho limits your drawing to up-down or left-right. You activate Ortho during drawing by typing the F8 function key before a segment you want constrained. F8 cycles through three modes of directional constraint: Off, Ortho, and Polar, which constrains drawing to 5-degree increments. When you are in Ortho or Polar mode, a compass appears at your cursor, and the Ortho or Polar buttons at the Status Line are active.

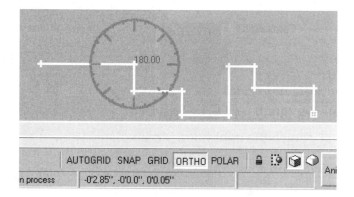

In AutoCAD you can click the Ortho button at the Status Line while you are in a Line command. You cannot do this in VIZ; the only way to cycle through Off/Ortho/Polar while drawing is with the F8 key.

188

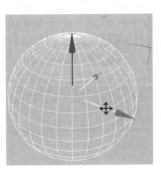

In AutoCAD, Ortho works while drawing and also while moving objects. In VIZ, Ortho only works while drawing. When transforming an object (Move, Rotate, Scale) in VIZ, you constrain direction by dragging one axis of the Transform Gizmo or by right-clicking and, from the pop-up menu, choosing Transform and then an axis constraint

Spline Editing

6. Once the basic shape is drawn, switch to the Modify panel, and turn on Sub-Object Vertex

7. Drag a selection window around the entire shape to select all the vertices

8. Type X to hide the Transform Gizmos

9. Right-click over any vertex, and from the pop-up menu choose Bezier Corner

When spline editing, should all the vertices always be set to Bezier Corner? Not necessarily. Setting them all to Bezier Corner is a quick, convenient way to begin editing the spline, and Bezier Corner allows the greatest flexibility at a vertex. But Bezier Corner also may result in a visible fold at that vertex in the 3D object, whereas the Bezier type (which keeps both tangent handles colinear) or the Smooth type (which has no tangent handles) assure smooth shading across that part of the object. For sharp angles without curvature, Corner is your best choice. The figure at right shows the best choices for the upper part of the candlestick shape.

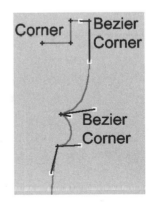

189

10. Click away from the Line to deselect all vertices

11. Turn off Ortho and Polar

12. Manipulate the vertices and tangent handles to get the right shape for the profile of the candlestick.

13. If there are excess vertices, select them and to delete them, either type the Delete key on the keyboard, or select the Delete button in the Modify panel. Remember, you want the minimum number of vertices required to achieve the desired shape

14. When you are satisfied with the profile, turn off Sub-Object

15. Name the Line %Candlestick Profile

16. Select the Rectangle you used as a guide and delete it

17. Select the Candlestick Profile

18. Type X to show the Transform Gizmo

19. Shift-Move the profile to one side to make a clone of it. In the Clone Options dialog box, choose Reference as the clone type, and name the clone Candlestick01. Choose OK

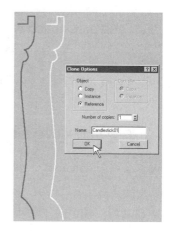

When you will be applying a modifier to a 2D shape, it is usually a good idea to make a Reference of the shape and apply the modifier to the Reference. Remember that a Reference is a child of the original; whatever you do to the original shape, you will also do to the Reference. But anything you do to the Reference (like applying a modifier) does not affect the original. So applying the modifier to the Reference allows you to spline edit the original shape and see the results updated immediately in the 3D object.

Apply the Lathe Modifier

1. With Candlestick01 selected, apply a Lathe Modifier

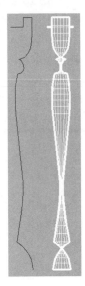

Sub-Object Axis

Looks interesting, but not like a candlestick yet. The axis of lathing sets itself by default to run through the centroid of the shape. Here, you need the axis to be positioned at the left side of the shape.

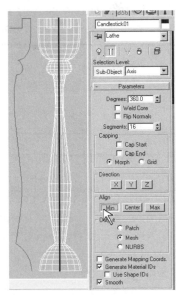

2. In the Lathe parameters, in the group labeled Align, choose the Min button. The object should look like a candlestick now

The default is for the axis to align to the center of the shape. If you are consistent in the way you draw profiles, always drawing the right profile of the object as you have done here, then the left side of the shape will be the Min button, and will result in the object you want. Also remember to always draw the profile in a counterclockwise direction, or your object will be flipped inside-out (if this happens, you may be able to fix it with the Flip Normals checkbox in the Lathe parameters).

191

3. At the bottom of the Modify panel, check Generate Mapping Coordinates

4. Uncheck Generate Material IDs

Generate Material IDs is on by default in the Lathe Modifier, and its effect is that the outside faces of the object get assigned material ID 3, which is an awkward choice. IDs 1 and 2 get assigned to faces that make up the cross-sections inside the object. So unless you want to lathe the object only partway around (less than 360 degrees) so as to reveal the cross-sections, uncheck Generate Material IDs. Then the object will carry ID 1, which makes material assignment easy. If you do want to reveal the cross-sections, the shape being lathed must be a closed shape, not an open one like this candlestick profile.

5. Type W to return to four views

6. Make the Perspective view active, Zoom Extents Selected to see the Candlestick close up, and type the F3 function key to set the view to Wireframe

7. In the Modify panel, find the Segments spinner and nudge its value up and down to see the effect. Try setting it all the way down to 3, then 4, then set it back to 16

Setting the Segments to 4 should suggest to you a wealth of possibilities for the Lathe modifier. It is the perfect modifier for newel posts, square column capitals, and square tabletops with ornate edges. A crown mold for an octagonal room is simple; draw the section, apply a Lathe modifier, set Segments to 8, then turn on Sub-Object Axis and move the axis of lathing to the center of the room.

8. Save the scene

Lathing a Picture Frame

Use a Line and a Lathe Modifier to model a picture frame.

1. Make the Top view active

2. Draw a small Rectangle, of no particular size, just behind the table

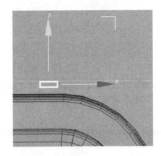

3. Set the Rectangle's parameters to:
    Length = 1.5″
    Width = 5″

4. Zoom Extents Selected

5. Using the Line tool, sketch a cross-section of a piece of molding, as seen from above. Keep it simple, using only ten to twelve vertices. Don't bother closing the shape at the back, since the back of the picture frame will be against the wall. Drawing counterclockwise means in this case that you will start at the upper-left corner of the Rectangular guide, and finish at the upper-right of the guide. Use Ortho for the straight segments

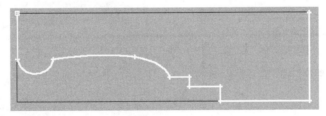

6. Edit the line in the Modify panel, so that it looks like the figure above

7. Name the line %Picture Frame Section

8. Delete the Rectangle guide

9. Shift-Move the cross-section to clone it. Choose Reference as the object type, and name the clone Picture Frame

10. Apply a Lathe modifier to the clone

11. In the Lathe parameters, set the Segments spinner to 4

12. Align the axis to the Min of the shape (the left side)

13. Check Generate Mapping Coordinates

14. Uncheck Generate Material IDs

15. Zoom Extents All Selected

At this point the object should look something like this:

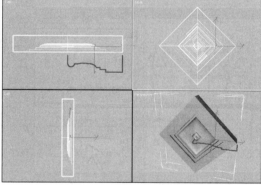

16. In the Modify panel, turn on Sub-Object

17. Choose the Move tool

18. In the Top view, position the cursor over the X axis of the Transform Gizmo, and move the axis of lathing to the left. As you do so, you will see the object open into a picture frame in the Front and Perspective views (the frame is in a diamond orientation at the moment). Zoom Extents All Selected to see the frame

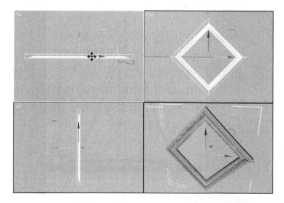

19. When you have opened up the frame as large as you want it, turn off Sub-Object

At this point, if you were not satisfied with the profile of the frame, you would only need to edit the 2D cross-section shape, and the 3D frame would update simultaneously

To finish the Picture Frame, Rotate it 45 degrees in the Front view, and add a canvas.

20. Make the Front view active, make sure the Sub-Object button is turned off, select the Rotate tool, then right-click over it to bring up the Transform Type-In

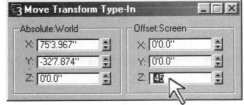

21. In the Offset: Screen Z: spinner, type -45, and hit Enter on the keyboard, then close the Transform Type-In

22. Zoom Extents, and move the frame above the table

The frame should now be squared up.

The canvas is just a Rectangle made into a plane with the UVW Map modifier.

23. In the Front view, Region Zoom tight to the picture frame

24. Activate Snap, then right-click over the Snap button and set snaps to Vertex

25. In the Front view, draw a Rectangle for the canvas, snapping to the inside corners of the frame

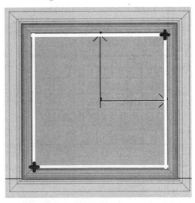

26. Turn off Snap

27. Name the Rectangle Canvas

28. In the Modify panel, apply a UVW Map modifier, to mesh the rectangle and prepare it to receive a bitmap

29. Save the scene

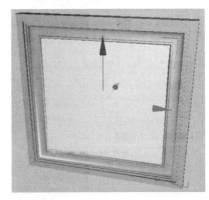

This technique for the picture frame is often the best way to model window and door trim. To make door trim, for example, lathe the cross-section into a picture frame, apply an Edit Mesh modifier, choose Sub-Object Vertex, and select all the vertices at the bottom of the frame. Then just move those vertices down and bury them in the floor. If you are a purist, you can remove the bottom faces and edit the vertices for a clean fit at the floor, but it is not really necessary.

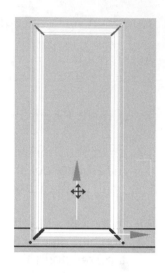

When faces interfere like this it is called occlusion. VIZ does not usually have problems rendering such a situation, but other programs may. In particular, if the model is to be exported to Lightscape, occlusion will create anomalies in the rendering, and must be avoided.

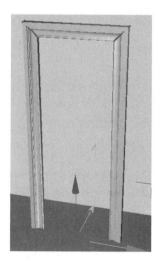

Other Lathe Parameters

Weld Core: In the Lathe parameters, use Weld Core to fix triangulation problems at the ends of lathed objects. For example, if you lathe a round table, you will likely see triangles radiating out across the tabletop if you don't Weld Core.

Capping: If the shape to be lathed is a closed loop, you can lathe less than 360 degrees, check capping, and cross-sections of the lathed object will be created. Morph Capping caps using the least possible number of triangles. Grid Capping caps using a grid of faces.

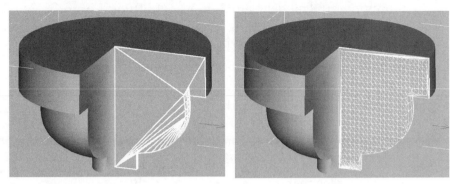

Morph Capping   Grid Capping

Direction: The Direction buttons rotate the axis of lathing. If you draw the cross-section to the right of the axis, and draw it in the correct view, you should not need to rotate the axis. If you import shapes to be lathed from another program, you may need to rotate the axis.

Cafe Chair: Lofting

A great variety of furnishings can be made with simple modifiers like Bevel, Bevel Profile, and Lathe. But when you need to model the curved back of a cafe chair, you'll need one of the more powerful modeling tools in VIZ: Lofting. Lofting is simple, conceptually; it is a shape extruded along a path. If you've modeled in AutoCAD you have likely used the Extrude command with the /Path option. That is lofting. You use lofting to model things like complex handrails, cornices, and crown molds in non square spaces, and many furniture parts, such as the legs and back of this chair that you will build in this section:

There are two features of lofting that make it particularly versatile. First, you can place different cross-section shapes at points along the path, as in a chair leg that is square at the floor, then oval where it joins the seat. Second, there is a set of Deformation tools you can apply to a Loft object to alter it, making it bulge, twist, taper, or conform to profiles anywhere along its length.

Draw Loft Shapes

You need several shapes with which to loft the chair. You need shapes in plan for the cross-sections of the legs (square at the floor, oval where they connect to the seat), and an oval for the shape of the seat. You need shapes in elevation for the paths of the front legs, the path of the back legs/backrest, the profile of the seat, and the shape of the wicker mesh in the seat back.

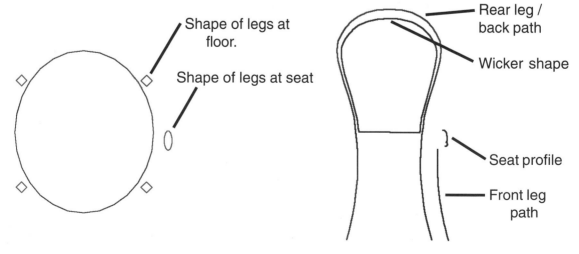

1. Make the Top view active

2. To the right of the pedestal and vase, draw a Rectangle of any size, then set its parameters to:

   Length = 14 ″
   Width = 16″

3. Zoom Extents Selected

4. In the Display Floater or the Display Panel, turn Off Unselected objects

5. Activate Snap, and set Snaps to Vertex

6. In the Create panel, Shapes category, choose NGon

7. Drag out a small NGon at the lower-right corner of the rectangle, snapping to the corner

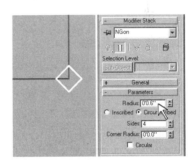

8. Set the Ngon parameters:

   Radius = 0.6″
   Sides = 4
   Corner Radius = 0

9. Name the NGon %Leg Square01

10. Hit the spacebar, or choose the padlock button below the views, to lock the selection

11. Use Shift-Move to make a duplicate of the NGon at the lower-left corner of the rectangle. Make sure you see the Snap marker showing before you begin to drag, and before you release at the other corner. In the Clone Options dialog box, choose Copy as the object type

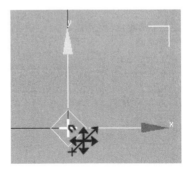

12. Repeat the Shift-Move to clone two more duplicate NGons at the upper-left corner and at the upper-right corner

13. Turn off Snap

14. To the right of the rectangle, drag out an Ellipse of any size, then set its parameters to:

> Length = 2.5″
> Width = 1″

15. Name the Ellipse %Leg Oval

16. Drag out another Ellipse of any size, then set its parameters to:

> Length = 21″
> Width = 18″

17. Use the Align tool to center the ellipse to the rectangle. In the Align dialog box, check the check boxes for X Position and Y Position, and set both Current Object and Target Object to Center

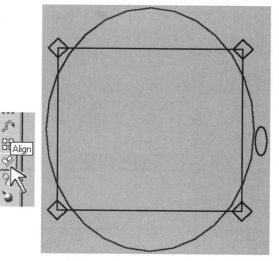

18. Name the Ellipse Seat

19. Select and delete the rectangle

20. Make the Front view active, then Pan and Zoom to see the shapes you just drew (viewed on-edge), and a few feet of space above them

21. In the Front view, draw a Rectangle of any size, then set its parameters to:

> Length = 44″
> Width = 20″

22. Zoom Extents Selected, and maximize the Front view

The placement of the rectangle (or any of the shapes you are about to draw in elevation) is not important. At lofting time, the paths will be moved to the location of the shapes you drew in plan view. When drawing shapes and paths for lofting, decide whether you want to move the shapes to the path (in which case you would draw the paths in their correct locations), or move the paths to the shapes (in which case you would draw the shapes in their correct location). Which you choose does not usually matter– do whatever is more convenient.

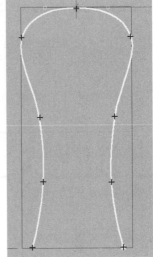

23. Using the Line tool, with the rectangle as a guide, draw the path for the piece that will be the back legs and back of the chair. Before you start to draw, remember to set Drag Type (in the Create panel) to Smooth. Start at the lower-right of the path, and draw counterclockwise. You should not need more than nine vertices to draw the path. You may be tempted to look for a way to draw half the path and mirror it, but at this point that may cause some problems, so just get it as symmetrical as possible by eye. Of course, as you draw it, don't worry about the exact shape– just get nine vertices in their approximate locations

24. Switch to the Modify panel, switch to Sub-Object Vertex, and edit the path. You will want to set the bottom two vertices to Corner type, the top vertex to Bezier type, and the rest can be Smooth type (which has no tangent handles). In the General rollout, set the Steps to 6

25. When the path is the right shape, turn off Sub-Object and name the path %Back Path

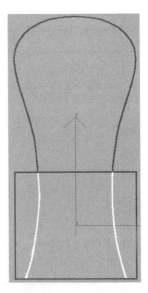

26. Change the Length of the rectangle to 18″, and move it down so the bottom of the rectangle is at the bottom of the path you just drew (by eye is close enough)

27. Use the Line tool twice to draw two paths for the front legs. Each path should have only three vertices. Use the % Back Path as a guide, and draw the two paths as tall as the rectangle. Draw both paths starting at the bottom of the rectangle. As seen in the Front view, the left-hand line should be named %Leg Path-L, and the right-hand line should be named %Leg Path-R

28. In the Modify panel, edit the two leg paths, if needed

29. Select and delete the Rectangle

30. Type H, select the two leg paths, and move them to one side a bit, so they are not right over the other path

31. Type W to return to four views, then Zoom Extents All

You should now have a set of shapes and paths set up something like this:

32. Save the scene

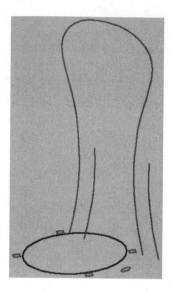

Create Loft Objects

Lofting is done in the Create panel. You are not altering the shapes and paths you've drawn, you are creating new objects based on the shapes and paths. The Loft object carries within it, as Sub-Objects, instances of both the shape and the path upon which the Loft is based, so that once the Loft object is made, you can select the shapes and paths, edit them, and the Loft object will immediately update.

1. Select %Leg Square01

2. Switch to the Create panel

3. In the Geometry category, open the subcategory drop-down list and choose Compound Objects

4. Choose object type Loft

5. In the Creation Method rollout, choose Get Path

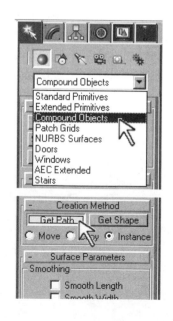

6. In any convenient view, select %Leg Path-R

7. In the Perspective view, Zoom Extents Selected

8. Switch to the Modify panel, and name the new object Leg01

9. Open the Skin Parameters rollout, and set the Shape Steps spinner to 3, and the Path Steps spinner to 6, to add more interpolations between vertices for smoother curves

10. In the Skin Parameters rollout uncheck Cap Start and Cap End

The Start is the surface of the leg that is against the floor. The End is the surface that will be against the bottom of the seat. Neither surface needs faces.

11. In the Surface Parameters rollout, check Smooth Length, Smooth Width, and Apply Mapping. If Generate Material IDs is checked, uncheck it

Without Smooth checked the rendered leg would show facets, either along the length of the leg, or around its sides. Without Apply Mapping checked, the renderer would not know how to apply the bitmap of wood that will be used in the material for the leg.

Notice the Length Repeat and Width Repeat spinners in the Mapping group. These values for a loft are important to remember. Suppose you were lofting a sidewalk- you would draw the sidewalk cross-section, then the path, then loft the section along the path. Then you would check Apply Mapping. The map that you would incorporate into the sidewalk material would be either a photo of some concrete sidewalk, or you could make a map yourself in a paint program. Ideally, that map would show a length of sidewalk with one expansion joint in the picture. Assuming that you want 8 feet between expansion joints in the sidewalk, and the sidewalk is 240 feet long, you would set the Length Repeat spinner to a value of 30 (because 30 x 8 = 240), and your lofted sidewalk would have perfectly spaced expansion joints.

The Loft object is bent in the opposite way from the path- the path bends to the right, the Loft bends left. Lofting is a bit unpredictable, and as often as not, the Loft object does not end up oriented the same way as the path. Fixing it is usually fairly simple:

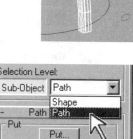

Rotate Loft Path

12. Choose Sub-Object. Open the Sub-Object drop-down list and choose Path

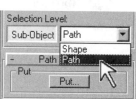

There is actually not that much you can do in Sub-Object Path mode of a Loft- there is only one button in the Modify panel, labeled Put. You would use this button if you had made a Loft earlier, deleted the original path upon which that Loft was based, and now wanted to edit that path. Put would put another instance of the path back in the scene, so you could edit it.

Above the views, look at the Transform buttons- Move, Rotate, and Scale. Only Rotate is available.

13. Choose the Rotate tool, and right-click over it to show the Transform Type-In

14. In the Transform Type-In, type 180 in the Offset:Local Z spinner, hit enter on the keyboard. The loft should now bend in the correct direction

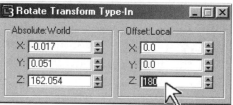

15. Turn off Sub-Object

16. Select %Leg Square02

17. Switch to the Create panel, and repeat steps 2 through 14 to loft a second chair leg. The path this time is %Leg Path-L. Name the new loft object Leg02

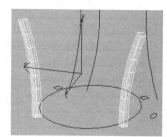

You have lofted two curved chair legs, and only need to rotate them a bit to get them in the correct position. Make all the various parts first, and then assemble them using transforms.

18. Select %Leg Square03

19. In the Create panel, choose Loft, choose Get Path, and click on the %Back Path line

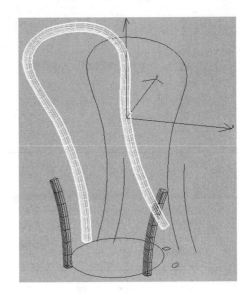

The back is created at an angle, but that's ok – the shape is correct.

20. In the Modify panel, Skin Parameters rollout, set Shape Steps to 3, and Path Steps to 10. Uncheck Cap Start and Cap End

21. In the Surface Parameters rollout, check Smooth Length, Smooth Width, and Apply Mapping, and uncheck Generate Material IDs

22. Name the new object Legs-Back, and save the scene.

The Seat: Bevel Profile

You will return to the Loft objects shortly to learn to change the cross-section of the loft at points along the path. But first, the seat is a perfect opportunity to revisit Bevel Profile. You have the oval shape, but you have not drawn the profile yet.

1. Make the Front view active

2. Draw a Rectangle, anywhere, any size, then set it to:

    Length = 3″
    Width = 1″

3. Zoom Extents Selected

4. With the Line tool, sketch the profile inside the rectangle, then edit the line to the proper shape in the Modify panel.

All three vertices should be drawn using Initial Type mouse-clicks, (straight segments), and the curvature at the vertices should be edited in. When editing the spline, open the General rollout and add as many Steps as are needed for reasonably smooth arcs.

5. Name the profile %Seat Profile

6. Delete the rectangle guide

7. Type H and Select Seat (the large oval)

8. In the Modify panel, apply a Bevel Profile Modifier (select More, and find it in the list)

9. In the Bevel Profile parameters, select the Pick Profile button, then in any convenient view, select the %Seat Profile

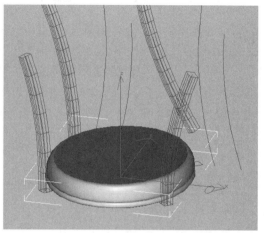

In a shaded view, you can see that the top of the seat has been assigned a different Material ID number than the sides. There is no check box for controlling this in the Bevel Profile Modifier. Before assigning materials, you will fix the Material ID numbers.

Transforms: Assemble the Chair

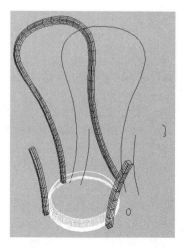

A couple quick Move and Rotate commands, and the chair will be assembled. Then you will add different shapes at points along the Loft paths to make the chair more interesting and realistic.

1. Make the Perspective view active, then Zoom Extents All, Zoom, and Arc-Rotate as needed, to get a good view of all the parts

2. With the seat still selected, select the Move tool, right-click over it, and in the Transform Type-In, type 18 in the Offset:World Z spinner, then hit Enter on the keyboard to lift the seat 18"

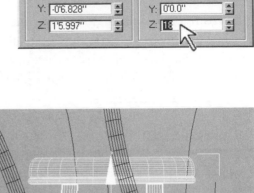

3. Select one of the front legs

4. Make the Front view active

5. Select the Rotate tool, and rotate the leg so that the top end of the leg will meet the bottom of the seat correctly

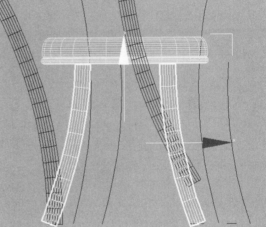

6. Move the leg into position. Look in the Top view to verify that the top of the leg is completely under the seat

7. Repeat the Rotate and Move for the other leg

8. Still in the Front view, select the seat back and rotate and move it to its correct position. Verify its position in all views

9. Render the Perspective view, and adjust the position of the parts if necessary

10. Save the scene

Multiple Shapes in a Loft

The three lofts you have made all have just a square as the cross-sectional shape. You can add as many additional shapes as you want along the loft path to change the object. The front legs of the chair will be square-shaped at the floor, and oval-shaped where they meet the bottom of the seat. The back legs/backrest will be square-shaped at the floor and oval-shaped at the top of the curve. The back should also be flattened and elongated at the top of the curve, for a place to support the sitter's back. Add oval cross-sections to the three lofts.

1. Select Leg01

2. In the Modify panel, open the Path Parameters rollout

The Path Parameters rollout is concerned with adding shapes along the path and with accessing shapes added earlier. The first decision in this rollout is how to measure how far along the path the next shape should be placed; either a certain Percentage of the path distance, or a given Distance, or just at the next Path Step (remember that Steps are interpolations between vertices, and there are usually six by default). Once this is decided, you enter a value for the percentage or distance in the spinner labeled Path, and a marker will travel this percentage or distance along the path, and wait for you to choose a shape to add. You scroll up the Modify panel, choose the same Get Shape button you used when first creating the Loft object, and click on the desired shape in the views. The Loft object will take on the new shape at the location of the marker.

3. Set the measurement choice to Percentage

4. In the Path spinner, type 100, and hit Enter on the keyboard

It is a bit hard to see, but if you zoom tight to the top of the leg, you'll see a yellow X at the top of the path that runs up the middle of the leg.

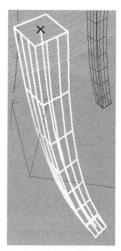

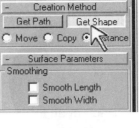

5. Near the top of the Modify panel, select the Get Shape button

6. In any convenient view, click on the small oval shape on the floor (%Leg Oval01)

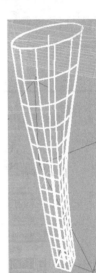

The leg now starts at the floor with a square shape, and ends at the seat with an oval shape. There is a smooth transition between the two shapes along the entire length of the leg. It might look better if the leg stayed square until about halfway up, and then began to change to oval.

7. Set the Path spinner to 50

8. The Get Shape button should still be active. Click on the square shape at the bottom of the leg

A Square cross-section is added halfway up the leg. There seems to be a new cluster of segments there now, right at the point of the new shape. This is due to a setting called Adaptive Path Steps, which is on by default. With Adaptive Path Steps on, anytime an additional shape is added to the loft, an additional vertex (and therefore additional steps) is also considered to be added at the point of the new shape, and so additional faces will appear. With Adaptive Path Steps off, the addition of new shapes is not considered to be adding any new vertices, and so no new steps and faces are added.

9. In the Skin Parameters rollout, uncheck Adaptive Path Steps

Rotate Loft Shapes

Now the change to oval starts halfway up. The rotation of the shapes in the leg could be improved. It would probably look better if the shapes were rotated so that the squares at the bottom were not in a diamond configuration, and the long axis of the oval at the top lined up with the curve of the seat at that point.

10. Turn on the Sub-Object button

11. In the Sub-Object drop-down list, choose Shape

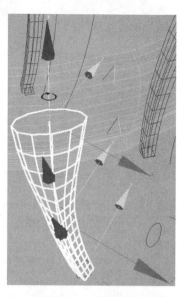

12. Drag a selection window around the entire leg to select all three shapes embedded within it

13. If the Transform Gizmos are not showing, type X to show them

14. Choose the Rotate transform button

15. Place the cursor over the Z axis of the topmost Transform Gizmo, as shown in this image, and drag upwards to rotate the shapes to square up the bottom and align the long axis of the oval at the top to the seat's curvature

As you drag, the surfaces of the leg will disappear, allowing a clear view of the three shapes as you rotate them. Refer to the Top view for correct rotation of the shapes.

16. Make the Perspective view active, get a close-up view of the leg, and render

The bottom of the leg does not meet the floor at the correct angle. This is easily fixed by rotating the bottom embedded shape in the Front view.

17. Make the Front view active

18. Region Zoom tight to the bottom of the leg

19. With Sub-Object still on, select just the square shape at the bottom and rotate it so the bottom of the leg is flat to the floor

20. Turn off Sub-Object

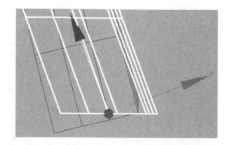

21. Select the other front leg and repeat steps 2 through 20 to add shapes and rotate the shapes

22. Save the scene

Now add a few shapes to the back legs / backrest loft. The backrest will have an oval cross-section at the top, and it should be a more elongated oval than the one used in the legs.

23. Shift-Move the oval shape that is on the floor (%Leg Oval) to make a clone of it. Choose Copy as the object type, and name the new oval %Backrest Oval

24. Edit the parameters of the new oval to:

    Length = 5″
    Width = 1″

25. Select the back legs / backrest loft, and Zoom Extents Selected

26. Add the long oval shape at 50% along the path (refer to steps 2 through 6 on the previous pages if you need help)

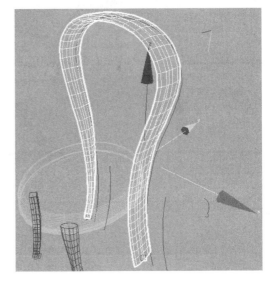

Now you have a Loft that starts out square, gradually changes to oval at the halfway point (the oval is not rotated correctly yet), and then stays oval-shaped to the end. You want it to return to square at the end.

27. In the Path Parameters rollout, set the Path spinner to 100

28. The Get Shape button should still be active. Click on one of the leg squares (it doesn't matter which one)

Now the loft returns to a square shape at its end. Like the front legs, it would look better if the legs stayed square for a distance before beginning to change to oval.

29. Set the Path spinner to 25.
    Get Shape should still be active.
    Click on the square shape again

30. Set the Path spinner to 75, and
    click on the square shape

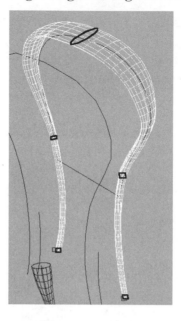

The Loft should look like this now:

The additional shapes are all in place.
Now they need to be rotated.

31. Turn on Sub-Object / Shape

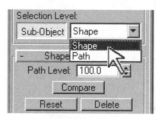

32. Drag a selection window around all the
    shapes in the Loft except the top one,
    the long oval

33. Select the Rotate tool, then right-click
    over it

34. In the Transform Type-In, type 45 in the
    Offset: Local Z spinner, hit Enter

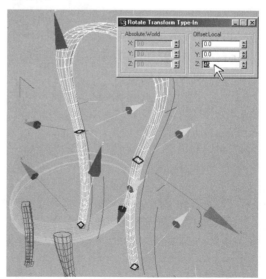

The bottom sections of the loft, the legs,
are now squared up, rather than in a
diamond orientation

35. Select the long oval shape at the top of the loft

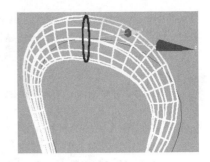

36. In the Transform Type-In, type 90 in the Offset: Local Z spinner, hit Enter

The backrest should now be more restful:

37. Make the Front view active and zoom tight to the bottoms of the back legs

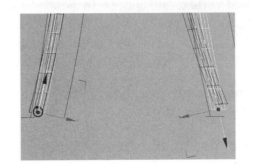

38. Select either square shape at the bottom of the legs

39. Rotate the square so the leg meets the floor properly

40. Repeat the last two steps for the other back leg

If the bottoms of the legs are not all at the same elevation off the floor after you have rotated all the shapes flat, getting them adequately aligned is a matter of using transforms on the three loft objects to get all the feet flat on the floor. In the image above, for example, the back legs need to be rotated clockwise just a fraction of a degree, and the right front leg needs to be moved down a bit. Of course, in the rendering, it is not likely that anyone will notice if a leg is an eighth of an inch too low or too high.

41. Turn off Sub-Object

42. Render a view of the entire chair

43. Save the scene

Two tasks remain before assigning materials: model the wicker in the backrest (a very easy task), and assign proper Material ID numbers to the seat.

1. Make the Front view active and zoom to see the entire backrest

2. Use the Line tool to trace a shape within the backrest, and ending just above the seat, as shown here. The exact shape isn't important; just trace an inch or two within the Loft. Make sure you click back over the first vertex, and choose to close the spline, and that you draw counter-clockwise

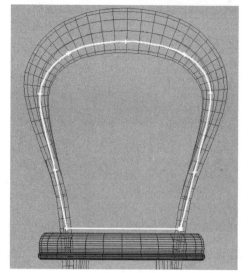

3. Switch to the Modify panel and edit the Spline to the correct shape

4. Apply a UVW Map Modifier to make the shape solid

5. In the parameters of the UVW Map modifier:

> Set Length = 10″
> Set Width = 10″

6. Name the shape Wicker

7. In the Top or Left view, Move the Wicker so that it is buried in the backrest. Render the Perspective view to verify the placement

Material ID Assignment

The profile you used in modeling the seat with the Bevel Profile modifier shows a trim bead at the bottom of the seat. The seat will carry Material ID 1 and the trim bead will carry Material ID 2.

1. Select the seat

2. Maximize the Front view and Zoom Extents Selected

3. Apply an Edit Mesh Modifier to the seat

4. Switch to Sub-Object Polygon

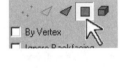

5. Set the Crossing / Window Selection button to Crossing Selection

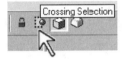

6. Drag a selection window, starting above the seat, to select all the faces that will be the seat material

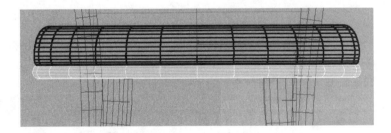

7. Scroll to the bottom of the Modify panel and set the Material ID spinner to 1

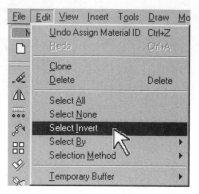

8. From the menus, choose Edit / Select Invert

9. Assign Material ID 2 to the selected faces – the trim bead

10. Turn off Sub-Object

The addition of the Edit Mesh modifier has caused the seat to lose its mapping coordinates. Replace them with a UVW Map modifier.

11. In the Modify panel, apply a UVW Map modifier

12. Set mapping Type to Box, then set the Length, Width, and Height all to 18"

13. Restore four views, Zoom Extents All

14. Save the scene

216

If you feel like it. . .

Bend Modifier

If the chair looks a bit stiff to you in the Left
view, and you have the energy, try a Bend
Modifier on the seat back. The proper values
for the Bend parameters are:.

> Angle = 22
> Direction = 90
> Bend Axis = Y

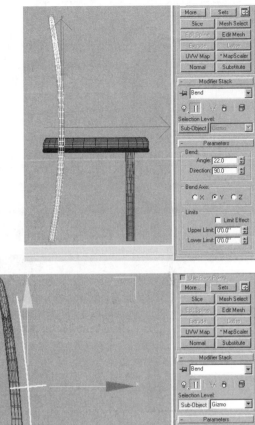

You'll also have to Bend the Wicker.
The parameters for that Bend are:

> Angle = 7
> Direction = -90 (note minus)
> Bend Axis = Y

After bending the Wicker, move and
rotate it to bury it in the backrest.

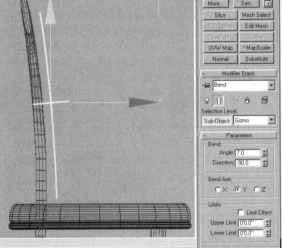

Bend the front legs if you want. It
takes a bit of trial-and-error to get the Bend parameters right. The best way to find the
proper Bend Axis (the axis you want to bend) is to set the Reference Coordinate System
drop-down list to Local, so you are certain that you're seeing
the object's own axes, and not the World axes, in all views.
Then look at the axis tripod in the views and decide which
axis needs to be bent. As for the Angle and Direction spinners, set the Angle to 20 or so, then try different increments
of 90 degrees in the Direction spinner and see how the object
behaves. When you have set the Bend parameters, set the
Reference Coordinate System back to View.

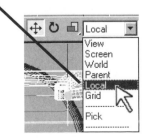

217

Materials

The legs and back of the chair will be wood, the seat a beige fabric with a metal trim bead, and there is the wicker backrest. There are suitable materials in libraries for all these objects.

1. Switch to four views and Zoom Extents All

2. Select the front legs and back legs / backrest

3. Open the Material Editor

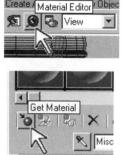

4. Choose Get Material

5. In the Material / Map Browser, in the Browse From group, select Mtl Library

6. In the File group, choose Open

7. Open Wood.mat

8. Drag material Wood-Bass from the Browser to an unused sample sphere in the Material Editor. If you only see six sample spheres, and they are all used (all show materials other than a default material), right-click over the active sample, and from the pop-up menu choose 5 x 3 Sample Windows

9. Choose Assign Material to Selection

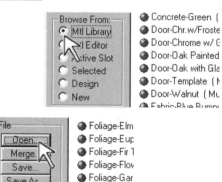

10. Select the Wicker

11. In the Material / Map Browser, open material library 3dsviz_big.mat

This is a large library and loads slowly– be patient. If you cannot open 3dsviz_big.mat at all, open the library Wicker.mat, saved in the C:\Viztutorials\Chapter4 folder.

12. Drag material Misc.-Wicker Leather to an unused sample sphere

13. Choose Assign Material To Selection

14. Make an unused sample sphere active

15. Select the button labeled Type, which is currently set to Standard.

16. From the Material / Map Browser, choose Multi / Sub-Object, choose OK

17. In the Replace Material dialog box, it doesn't matter whether you choose to discard or keep the old material, because you have not worked on the material. Leave it at the default and choose OK

18. Select the Set Number button, and set the number of submaterials to 2

19. Name the topmost level of this material Seat

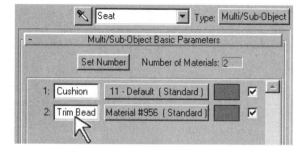

20. Name the two submaterials Cushion and Trim Bead

21. In the Material / Map Browser, open material library Fabric.mat (if the Browser is not open, open it with the Get Material button)

22. From the Fabric.mat library, drag Fabric-Beige onto the bar for submaterial Cushion

23. Open material library 3dsviz.mat

24. Drag material Metal-Brass onto the bar for submaterial Trim Bead

25. Select the seat

26. Choose the Assign Material to Selection button

27. Close the Material / Map Browser and the Material Editor, and save the scene

28. Get a good view of the entire chair in the Perspective view, and render

If you get a warning dialog box stating that any object is missing Mapping Coordinates, cancel the rendering, select the object in question, and examine the Modify panel for that object. Is there a Generate Mapping Coordinates check box or an Apply Mapping check box that is not checked? If so, check it and re-render. If the last modifier that was applied to the object is a Bend modifier, open the Modifier Stack, which reads Bend, and drop down to the Loft level to find the Apply Mapping check box.

Not a bad-looking chair. It might not suit your taste in furnishings, but it is well-modeled.

Group the Chair

When you merge this chair into other scenes, you want the entire chair to be selectable as one item, not as the five pieces that make it up. Group the chair.

1. Select all the parts of the chair (just the 3D parts, not the 2D shapes)

2. From the menus at the top of the interface, choose Draw / Group / Create

3. Name the Group Cafe Chair01

Additional Materials

You know enough now to assign premade materials from the various libraries to the furnishings that do not have materials yet. The only item you will not find a premade material for is the canvas in the picture. You will probably need a bit more knowledge about materials and mapping to apply an image to the canvas, and you'll learn that in the last chapter.

Before you save this scene and put it away, show all the furnishings and render them.

1. Use the Display Floater (Tools / Floaters / Display Floater) or the Display panel to show all 3D furnishings, and hide all 2D shapes

2. Arrange the furnishings next to each other

3. Save the scene

4. Render the set of furnishings

Summary

If your scene looks similar to the image on the previous page, then congratulations. This was a long chapter and a lot of work. Once you've practiced the techniques of this chapter in modeling some furnishings of your own, you will have enough skills to model just about anything in an architectural visualization except a landscape. You know how to build walls, doors, windows, stairs, and handrails. If you want a fancy handrail for the stairs, you know how to draw the cross-section and loft it up the railing path. If you need more realistic trim than what the parametric door and window objects offer, you can turn off Create Frame for those objects, draw a cross-section of the trim you want, and use Lathe with the picture-frame technique to make new trim. How would you model a chair-rail traveling around a room? Trace the shape of the room with a line, draw the cross-section of the chair rail with another line, and loft the section along the path. How about the fascia and drip-edge of a roofline, traveling up and down gables? That is also done with lofting. The only things you cannot model with these basic tools are organic forms, and objects made of complex-curved, joined surfaces, like a car body.

Earlier it was said that you do not need a huge number of tools to model buildings, but rather a solid knowledge of a few powerful tools. This is your tool kit so far:
• Primitives
• Wall, Door, Window, Stair, Railing parametric objects
• Editable Spline, or the Edit Spline modifier
• Editable Mesh, or the Edit Mesh modifier
• Meshsmooth modifier
• Bevel modifier, Bevel Profile modifier
• Lathe modifier
• Lofting

For modeling buildings, there are a few more modifiers that are often useful. The Lattice modifier is for making struts with joints- perfect for building spaceframes. The PathDeform modifier makes an object bend to follow a path- use it for lettering over an arched entryway. You will find occasional use for basic modifiers like Bend, Taper, and Twist.

You would be wise to spend some time practicing and experimenting on your own before starting the next chapter, to imprint skills in your long-term memory and to gain comfort and confidence before moving into new subjects. Working through tutorials is a good way to gain information, but to retain it you have to make simple scenes yourself.

223

CHAPTER 5

Landscapes and Natural Materials

Landscapes are usually the most challenging part of any visualization project. They are easy if the terrain is fairly flat, but a hilly site with numerous topographical features such as roads, curbs, retaining walls, planting areas, water, and building foundations set into the hills may actually require more work than the buildings themselves. Plan for this when preparing a proposal for a visualization involving a feature-rich landscape.

Landscapes usually begin as 2D linework in a drafting program. In this exercise you will import an AutoCAD 14 file as the basis for the terrain. If your drafting program is AutoCAD, you may be tempted to use the DWG Link feature to attach the 2D linework to the VIZ scene, with the idea that as you make changes to the landscape design in AutoCAD, those changes will be almost automatically updated in VIZ. This is a great concept, and one which can work well for other elements of your scene, but when it comes to terrain, it is simply asking too much of the software. Before starting to model a complex landscape, be confident that the design is fairly complete, and if you know that the design will change, accept (and budget for) the fact that the changes may require you to remodel the terrain from scratch.

Preparing the CAD File

Whichever CAD program you use to draft the linework for the topography lines and the topographical features, adhere to the following guidelines:

The topography lines should be located at their correct Z value. In the rare case that you are importing a file from a drafting package that you do not own or don't know how to use, the topo lines can be put at their proper elevations in VIZ, but it is easier if done in the drafting program.

Shapes for roads, walks, retaining walls, and such do not need to be at the correct elevation. Just leave them at $Z=0$.

It is usually best if all the topo lines are on the same layer. Since this is probably not the best situation for a construction document, you will need to save a version of the CAD file set up for export to VIZ.

The linework representing landscape features may need to be divided up among extra layers, rather than having all the sidewalk shapes, for example, on one layer. The shapes of sidewalks, roads, planter beds, and such will be projected down through the fabric of the terrain to define the features. It is like cutting shapes in dough with a cookie cutter (except the dough here is not flat). If the cookie cutter shape is too complex, the computer is overwhelmed, and locks up. So if there are a dozen sidewalk shapes to be cut into the terrain, you should probably make several layers among which to distribute sidewalk shapes, so that no single layer's contents are too complex.

If your drafting program is AutoCAD, you can draw topos with Plines or with Splines (or with a combination of the two).

Importing the CAD File

This is the AutoCAD file you will import to build the landscape. The topo lines and other shapes were drawn using a combination of Plines and Splines. All the topos are on the same layer, and each has the appropriate Z value. Each topo represents four feet of elevation. There are shapes defining a driveway and parking area, a pond, two areas of trees to be planted, and the slab for a large house. Each shape is on its own layer.

1. In a new, empty VIZ scene, choose, from the menus, Insert / AutoCAD .DWG (do not choose Linked .DWG)

2. In the import dialog box, browse to the C:\Viztutorials\Chapter5 folder, select HillHouseSite.dwg to be imported, choose Open

3. Leave the DWG Import dialog box set to Merge objects with current design, and choose OK

4. In the Import AutoCAD DWG File preferences dialog box, leave everything at the defaults except for ACIS Options / Surface Deviation. Set the Deviation to 1.0, then choose OK

This value determines how certain curved AutoCAD objects (including curved Solids and Splines) will be converted to faceted objects in VIZ. A curved segment will be redefined with short facets, and the deviation value sets how far from the curve a facet is allowed to deviate before it breaks and a new facet begins. The default is .1, which is a very small value for an architectural project (it is an appropriate value for importing a model of a machine part). A value of 1.0 seems to work well for most architectural models.

227

This figure shows AutoCAD Splines imported first with Surface Deviation set to the default of .1, which results in far more vertices than necessary, and then with Surface Deviation set to 1.0, which results in fewer vertices, but still well-defined shapes. In fact, these topos could be imported with a higher Surface Deviation value and still retain their shape.

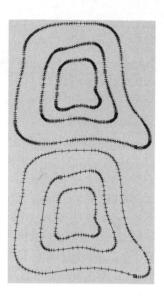

Occasionally an AutoCAD file will contain curved Splines or Solids that should not all be imported with the same Surface Deviation. You might freeze layers containing more detailed objects with tighter curves, import at a Deviation value of 1.0, then freeze the larger curved objects, thaw the detail items, and import again with a Deviation of .1 to add needed definition.

Normalize Spline Modifier

The first thing to do to the topos before converting them to a Terrain object is to fine-tune the number and spacing of vertices in the splines. This is done with the Normalize Spline modifier, which simply replaces all the vertices in a spline with new ones exactly spaced at a distance that you set. If you convert the topos to a Terrain object without using Normalize Spline first, there will be areas of the Terrain with small triangles bunched together, and areas with large triangles making big flat planes. Normalize Spline has only one parameter, Segment Length. The default for this value is 20, which means a vertex will be placed every twenty inches along the spline. For a spline defining a table-top this is fine, but for a landscape of several hundred feet, that is a huge number of vertices, and it can take several minutes to place the vertices, or the computer may lock up. The solution is to draw a small, simple shape, apply a Normalize Spline modifier to it, and set the Segment Length value to a larger number, say 150 (inches). This sets the default for your current editing session to 150, and when you then apply a Normalize Spline modifier to a large, complex spline like a topo map, it will generate the new vertices quickly.

1. Select the topos, and switch to the Modify panel

2. Switch to Sub-Object Vertex

You can see that the vertices are not evenly spaced, and that there are more than are needed.

228

3. Turn off Sub-Object

4. Switch to the Create panel, category Shapes

5. Off to one side of the topos, draw a small Circle of no particular size

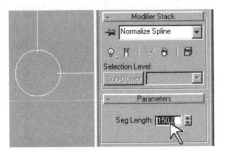

6. Switch to the Modify panel, choose the More button to see the list of additional modifiers, and choose Normalize Spline

7. Set the Seg Length spinner to 150

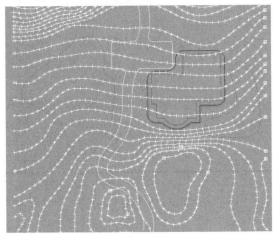

8. Select the topos and apply a Normalize Spline modifier to them

9. Apply an Edit Spline modifier to the topos, and switch to Sub-Object Vertex

The vertices of the topos are now evenly spaced, and there are far fewer of them.

10. Turn off Sub-Object

11. Delete the Circle, then select the topos again

Terrain Object

The Terrain object is made in the Create panel- you are not applying a modifier to the topo splines, you are converting them to a wholly new object.

1. Switch to the Create panel, Geometry category

2. Open the subcategory drop-down list and choose AEC Extondod

3. Choose Terrain

At this point, you may need to be patient for a few moments while the computer calculates the mesh. On a Pentium2-266 with 256 MB RAM, the Terrain appears in only a couple seconds. But this is a very simple set of topo lines, and in actual practice, with a complex set of topos, it can take up to a couple minutes to generate the mesh.

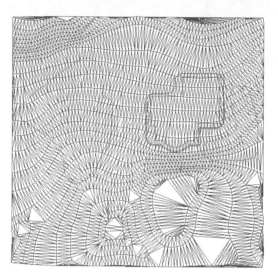

4. Once the Terrain appears, save the scene as Hillysite.max

Terrain Object Parameters

Though it looks like simply a group of irregular triangles forming a mesh, this Terrain object is a versatile parametric object, at least for now. When you begin to project shapes down through it to define areas of road, building foundation, planter beds, and such, then it will lose its parametric properties and become a simple Editable Mesh.

Operands: Operands are the splines that are the skeleton of the Terrain. This Terrain has only one operand, because all the topo lines were on the same AutoCAD layer. This is usually the simplest way to arrange things, but you may choose to import a CAD file in which the topos are on many layers. In this case, you would select the lowest topo line(s), convert that selection to a Terrain object, choose the Pick Operand button, select the next highest set of topos, then the next, and so on until all the sets of topo lines needed to comprise the finished Terrain are added as Operands.

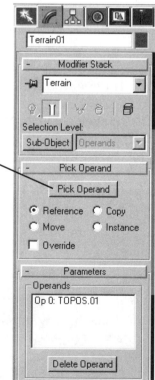

The most common scenario for addition of more Operands to a Terrain is when you see that certain areas lack sufficient density of triangles to smoothly accomplish a steep change in elevation. You can draw interpolating Splines in VIZ, put them at their correct Z height, and add them as Operands of the Terrain, giving more mesh density where it is needed.

230

Form:

Graded Surface: The default. A fabric of triangles.

Graded Solid: A skirtboard and underside are added.

Layered Solid: The terraced look of a chipboard model.

Display:

Choose to see only the mesh, or only the Operands (contour lines), or both, for an interesting presentation technique.

Update: Determines how changes in parameters are updated in the mesh: immediately (Always), only in the rendered image and not in the views (When Rendering), or only when you choose the Update button (Manually).

Simplification:

Horizontal Simplification allows the use of only 1/2 or 1/4 of the vertices in generating triangles, or the interpolation of vertices for a denser mesh. This is better controlled with a Normalize Spline modifier, for a more regular mesh.

If the CAD file contained more topo lines than necessary, resulting in an overly dense mesh, Vertical Simplification uses only 1/2 or 1/4 of the topos to generate the triangular mesh.

Color By Elevation:

Assigns layered zones of colors to the Terrain, according to elevation. This is an opportunity to instantaneously convey the topographical qualities of a landscape in a representational way, without needing to create a material or address mapping instructions. The Create Defaults button creates five or six zones of color, with blues representing water level or underwater, dark yellow representing shoreline, greens representing the middle elevations, pale yellows the higher elevations, and light gray representing the highest points in the landscape.

You can alter the elevation and the color of each zone by highlighting a zone in the list, changing the Base Elev spinner or the Base Color swatch, then choosing the Modify Zone button.

231

Terrain: Form

1. Make the Left view active, Arc-Rotate to get a good User (axon) view of the Terrain, then type the F3 key to set the view to shaded mode

2. In the Modify panel, in the group labeled Form, choose Graded Solid

3. Arc-Rotate in the User view again and examine the Terrain from the sides and from below. It now has thickness to the edges, and a solid face underneath

4. Choose Layered Solid, and Arc-Rotate to examine the Terrain. It is terraced now, and looks like a chipboard model

5. Set Form back to Graded Solid

6. Check Stitch Border – you will see a subtle difference at the edges. Uncheck it

Think of the heavy stitching added to reinforce the frayed edges of a rug– this is similar, but it almost never results in a better-looking model.

7. Set the view back to Wireframe mode

8. Check Retriangulate – a few triangles near the pond subdivide. Uncheck it

Retriangulate follows the original contours more faithfully when generating triangles. Use it only if you need it, as it makes the Terrain respond to editing more slowly. It is used most commonly to correct problems caused when a contour makes a sharp turn.

Terrain: Display

1. Set the User view back to shaded mode

2. In the Display group, first choose Contours, then choose Both, then set it back to Terrain

Terrain: Update

1. Set the Update choice to Manually.

2. Set the Form of the Terrain to Layered Solid. You won't see any change in the view

232

3. Choose the Update button - the Terrain displays as terraced

4. Set the Update choice back to Always

5. Set the Form back to Graded Solid

The manual update is useful on large Terrain meshes. It allows you to set several parameters without having to wait for the mesh to update after each parameter change. If you are performing a series of renderings to test the effects of different parameters, then When Rendering is the best choice. Like Manually, it will not update the mesh in the view until you select the Update button, but it will update parameter changes in the rendering each time you re-render.

Terrain: Simplification

1. Open the Simplification rollout

2. Set Horizontal Simplification to Use 1/2 of Points

3. Use the F3 function key to toggle between shaded and wireframe modes, observing the effect of using 1/2 of Points versus No Simplification in both modes

4. View the other choices for simplification and interpolation in both wireframe and shaded modes

5. Set Horizontal Simplification back to No Simplification

6. View the results of Vertical Simplification in both wireframe and shaded modes

7. Set Vertical Simplification back to No Simplification, and set the display mode back to shaded mode

Again, control of the number of horizontal facets is better done with Normalize Spline. Vertical simplification is probably also better done selectively, by adding or removing topo lines (in the CAD file or right in VIZ) only where needed, rather than allowing Vertical Simplification to affect the entire Terrain

Terrain: Color by Elevation

1. Open the Color by Elevation rollout

2. Choose the button labeled Create Defaults

In the CAD file, the topo representing the surface of the pond was at elevation Z=0, so Color by Elevation paints that level, and anything below it, blue. The surface of a body of water does not have to be at Z=0 in the CAD file for this to work. Suppose the landscape features a lake that in reality is at elevation 425 feet above sea level, and the topos representing the lake are at that Z value in the CAD file. In the Color by Elevation parameters, setting the Reference Elev spinner to 425′ would cause the lake surface to be shown blue. Of course, any part of the terrain with elevation below the lake surface would also turn blue, so there are limitations to what can be depicted with Color by Elevation.

Combining Color by Elevation with the display of both the Terrain and the Contours offers a quick way of conveying the topography in a clear, representational way.

By altering the colors of each zone, setting the lowest zones to black and the highest zones to white, a grayscale image showing the landscape's topography can be rendered from directly above. This image can be highly useful later, providing a means of placing and blending various natural materials, and also serving as a Displacement Map, in which the brightness of each pixel in an image is used to sculpt a mesh, with darker pixels pushing the mesh down and brighter pixels raising the mesh. It is another way of modeling a landscape, and you will try this at the end of this chapter. So now is the time to render the Displacement Map.

3. In the white list box that shows the elevations of the six color zones, highlight the first zone in the list

4. Below the list, in the Color Zone group, set the Base Elev spinner to 0

5. Choose the Base Color swatch to open the Color Selector

6. Set the color to H = 0, S = 0, V = 0

234

7. Choose the button labeled Modify Zone

8. Highlight another elevation in the list.

It doesn't matter which elevation you highlight because as you modify zones, the list gets rearranged. So as you work, just pick any zone that you have not modified yet, and when you're done, the list will be rearranged with the six elevations in their correct ascending order.

9. Set the Base Elev spinner to 15′, set the color selector to H = 0, S = 0, V = 50, then choose the Modify Zone button

This landscape's elevation varies from minus 4′ for the floor of the pond to 88′ at the top of the highest hill. You will place color zones at every 15′ of elevation, and each zone will increase in its brightness value by 50 points.

10. Highlight another unmodified elevation, set the Base Elev spinner to 30′, set the Color Selector to H = 0, S = 0, V = 100, then choose the Modify Zone button

11. Continue modifying zones until the list is set up like this:
    Zone at 0′   Value = 0
    Zone at 15′   Value = 50
    Zone at 30′   Value = 100
    Zone at 45′   Value = 150
    Zone at 60′   Value = 200
    Zone at 75′   Value = 250

12. Save the scene and close the Color Selector

13. Make the Top view active, and Zoom Extents Selected so the Terrain fills the view

14. Choose the leftmost teapot to open the Render Scene dialog box

15. Set the Width spinner to 2048, and set the Height spinner to 1536

This is a moderately high resolution for a bitmap. Displacement maps need to be high-resolution to allow accurate sculpting of landscape features. This particular map actually does not need to be this large, since this terrain is a simple one, but the more detail there is in a landscape, the larger the displacement map will need to be to sculpt accurately.

16. In the Render Output group, choose the button labeled Files

17. In the Render Output File dialog box, browse to the C:\Viztutorials\Chapter5 folder, name the file HillsDisplace, open the Save As Type dropdown list and choose .TGA as the file type, then choose Save. In the Targa Image Control dialog box, set Bits-Per-Pixel to 32, leave everything else as is, then choose OK

18. In the Render Design dialog box, choose Render

The displacement map should look like the image at right. The black background will need to be cropped out for this bitmap to be used as a displacement map, otherwise the black pixels might find the very edges of the mesh that the displacement map is applied to, and these edges would be pulled down to the zero elevation. The cropping can be done in a paint program, but VIZ also has a means of cropping a bitmap.

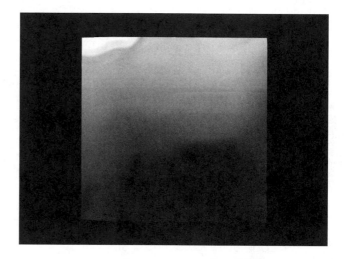

19. Close the rendered view, and close the Render Design dialog box

Incremental Saves

If you always save your work under the same file name, hour after hour, day after day, you will certainly pay the price for this eventually. Two things go wrong when you save to the same file each time. One is that you'll realize that earlier you accidentally modified or deleted an object, did not notice it, and saved over the file, making that object unretrievable. The other mishap is file corruption, which does occur with VIZ files. Sooner or later you will encounter a scene that was worked on the day before and correctly saved, but for some reason it simply refuses to open. You can try starting an empty scene and Merging the objects in the corrupt file into the new scene, but sometimes even that won't work. As a last resort, try opening the file on a different computer, if another one with VIZ installed is available. If none of these strategies work, and if that is the only name under which you've been saving that scene, you are out of luck, and you'll have to start over. The answer is to incremental save, frequently. The first time you save the file, name it House01. Work for a while, saving frequently, then use File / Save As to save as House02. By the time the project is finished, you may be saving as House30. How often should you incremental save? This depends on how much work you are willing to lose if a catastrophy occurs. Some people incremental save at every save. This may be overkill, but if after your lunch break you're still working and saving under the same name as when you began your day, you are begging for trouble. This topic is presented here because Terrain objects seem to be particularly prone to corruption, even when they are no longer parametric objects (they are collapsed to Editable Mesh). When modeling a large, complex landscape, incremental save every time. There is a handy button that makes this an easy habit to develop.

1. From the menus, choose File / Save As

2. In the Save File As dialog box, choose the button labeled with a plus sign

The scene is now saved as Hillysite01.max

Shapemerge-Cutting In Features

The features of the landscape now need to be defined in the Terrain. Each shape that represents a landscape feature will be projected through the Terrain with Shapemerge. The Terrain will be divided along the shape, and the faces within the shape will be easily selectable, so they can be assigned a unique Material ID number. Once all the shapes have been Shapemerged, and all the new areas assigned Material IDs, a Multi/Sub-Object material incorporating submaterials for grass, road, and mulch will be assigned.

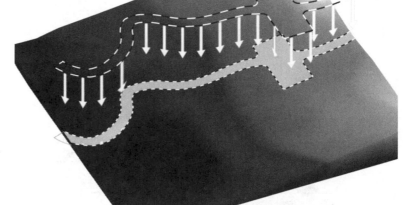

1. With the Terrain still selected, switch to the Create panel

2. In the Geometry category, open the subcategory drop-down list and choose Compound Objects

You are not about to modify the Terrain object, but rather you are going to replace the parametric Terrain object with a completely new compound object, a Shapemerge object. It is called a compound object because it is the result of the combination of two other objects: the Terrain, and the spline defining the shape of the road. Those two objects will remain embedded as Sub-Objects (Operands) of the Shapemerge, so that if the road were not quite in the right place, you could turn on Sub-Object, select the road shape operand, and move it around. The slice through the landscape would follow the movement of the shape. While this ability is impressive, it is not something you are likely to do in a large landscape model, as it may overwhelm the computer.

3. Choose the Shapemerge button

4. In the Shapemerge parameters, choose the Pick Shape button

5. In the Top view, position the cursor over the yellow road shape, called Drive_Parking01, and click

At first, nothing will appear to have happened. A few moments after selecting the shape, you might see the mesh change slightly as the shape cuts through it and new faces and edges are defined. To tell if it has happened, move the cursor around over the Create panel; if it is responsive, changing appearance and showing tooltips, the Shapemerge has been performed. For reference, the merging of the road shape takes 15 seconds on a Pentium2-266 with 256 MB RAM. It is not uncommon for a merge to take a minute or more. When merging a complex shape, or merging into a large terrain model, be patient, and wait a couple minutes before deciding that the computer has locked up.

If the shape you are trying to Shapemerge does overwhelm the computer, you will need to find a way to break the shape down into smaller shapes. You can redraft the shapes in your CAD program, looking for ways to divide large, complex shapes into smaller, simpler ones, then distribute the smaller shapes over a few extra layers, then import them and try again.

Mesh Select Modifier

Now that the road shape is defined in the new Shapemerge object, you need to select just the faces of the road and assign them a unique Material ID number.

1. Switch to the Modify panel

2. Apply a Mesh Select modifier to the Shapemerge object

3. Choose Sub-Object Polygon mode

The faces defined as the road surface become selected. That's all Mesh Select does – it just holds a selection of faces (or vertices or edges) so that another modifier can be applied to that selection

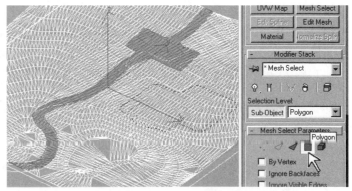

239

Material Modifier

4. Still in Sub-Object mode, apply a Material modifier

5. Set the Material ID spinner to 2

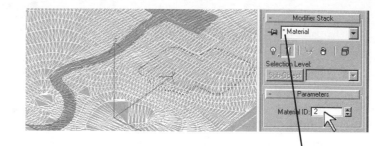

Look above the Material parameters at the Modifier Stack drop-down list—it reads Material, and notice that there is a small star before the name of the modifier. The star tells you that the modifier was not applied to the whole object, but to whatever Sub-Object selection was active when the modifier was assigned.

6. Incremental Save the scene

Collapsing the Stack

1. Select anywhere on the Modifier Stack to open it

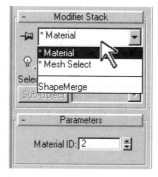

This is the history of the object. It was a Shapemerge object, then you added a Mesh Select modifier, turned on Sub-Object, and added a Material modifier in Sub-Object mode. You are free to move to any level in the Modifier Stack to change the parameters of any modifier in the stack. This is a powerful feature of VIZ. It keeps objects flexible, easily edited, and it makes your decisions less permanent. But you pay a price for this flexibility. The recording of each modifier in the stack takes memory. With numerous modifiers stacked up, a simple object might take on a fairly large file size. To display an object in the views, VIZ must begin at the bottom of the stack, calculate the effect of every modifier up the stack to arrive at the shape of the object, and then process the object's Transforms (its location in 3D space, its rotation, and its scaling) to finally place the object in the scene to be viewed through the viewports. When many objects in the scene are carrying many modifiers, things can slow down tremendously.

If you decide that access to all the modifiers is no longer necessary, you can "collapse the stack," to reduce the object to its simplest form, an Editable Mesh.

2. Click on the Modifier Stack drop-down again to close it

3. Below the Modifier Stack are a few icons, and one, showing a filing cabinet, is called Edit Stack. Choose it

4. In the Edit Modifier Stack dialog box, choose the button near the lower-right labeled Collapse All

A warning appears telling you that collapsing the stack will reduce the object to its simplest, nonparametric form, and any modifiers will no longer be present for editing. Once you understand the implications of collapsing the stack, you will probably want to check the check box at the bottom of this dialog box that says "Do not show this message again."

5. In the warning dialog box, choose Yes

The Modifier Stack now shows only one entry, Editable Mesh.

6. Choose OK to close the Edit Modifier Stack dialog box

Shapemerge the Tree Areas

1. Return to the Create panel

2. Choose the Shapemerge button

3. Choose the Pick Shape button

The shape defining the tree areas is called TREEAREAS.01. It is selectable in the views, but it is a bit difficult to see. If you aren't able to get a clear shot at picking on an object in the views, picking it from a list is always the safest way to go.

4. Type H, and in the Pick Object dialog box, highlight TREEAREAS.01, then choose Pick

Again, the shape will take a few moments to be merged through the terrain. This time it is easy to know when the process is complete- the Pick Object dialog box disappears.

Mesh Select, Material Modifiers

1. Switch to the Modify panel

2. Apply a Mesh Select modifier

3. Switch to Sub-Object Polygon

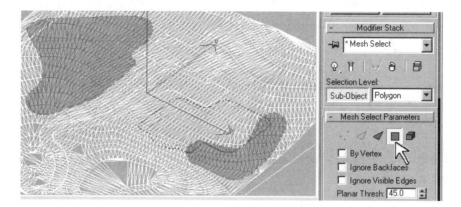

The faces comprising the two areas to be planted with trees are selected.

4. Apply a Material modifier

5. Set the Material ID spinner to 3

Collapse The Stack

1. Repeat steps 3 through 6 on the previous page to collapse the stack again

At this point the terrain is again an Editable Mesh, with areas of faces within it carrying three Material ID numbers, the first for grass areas, the second for the road, and the third for areas of trees.

Verify Material IDs

You do not want to continue developing this landscape if things are not correct so far, so take a moment to verify what you have done.

1. In the Modify panel, switch to Sub-Object Polygon

2. Scroll to near the bottom of the panel, to the Surface Properties rollout

3. Choose the button labeled Select By ID

4. In the Select By Material ID dialog box, verify that the spinner is set to 1, then choose OK

All the faces except for the road surface and the tree areas should highlight.

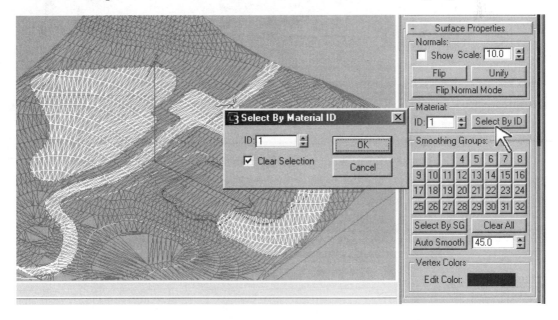

5. Choose the Select By ID button again, set the spinner to 2, and choose OK. The road should be highlighted. Select By ID again, using ID 3, and verify that the tree areas are selected

6. Scroll back up the Modify panel and turn off Sub-Object

If the Select By ID process showed any problems with the Material ID assignment, work back through the last few pages to correct the problem.

The Building Foundation

The site where the house will sit is not flat yet. The house is set into the slope, so that a retaining wall is needed at the front (the downhill side) of the house, and some excavation is needed at the back. There is a tool in the Terrain object parameters called Override, which allows the addition of another shape (spline) as an operand of the Terrain object, and the Override function creates a flat plane within the new shape. Any topos passing through the shape are ignored. Unfortunately, the Override function just doesn't work well in many situations. If the area you are making planar is hilly, then Override

results in rapid changes in elevation over short distances, and the function does not interpolate any additional faces at those points to help make the transition smoothly. The result is a very choppy terrain near the plateau, as you can see in the figure at right, which shows the results of using Override at the foundation site.

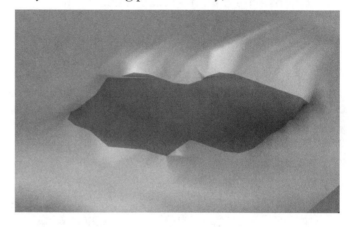

If your site is fairly flat, you might try using Override to generate slabs for your buildings, but for more hilly sites, this task involves a bit more work. If you do not need to show the earth excavated away to accept the building form, then don't bother- just bury the building into the hill, show some plantings about the foundation to hide the razorline junction of earth and building, and be done with it. When the design calls for a building with a terraced garden around it, the procedure is to cut the shape of the terrace out entirely, model the retaining walls at the front and back of the terrace, and model a flat floor for the terrace.

Shapemerge- Cookie Cutter

The two previous Shapemerge operations left the faces embedded in the terrain. The Cookie-Cutter setting in Shapemerge removes the faces inside the shape.

1. Switch to the Create panel

2. Choose the Shapemerge button

244

3. Scroll the Create panel to the group labeled Operation, and set it to Cookie Cutter

4. Choose the Pick Shape button

5. Type H, and from the Pick Object dialog box, highlight SLAB.01, then choose Pick

The shape of the terrace slab will be removed from the terrain.

6. Once the shape is removed, turn off the Pick Shape button

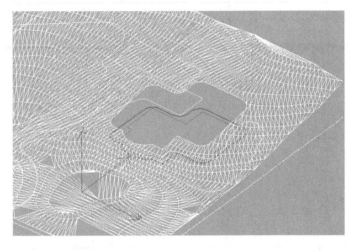

Mesh Editing – Extrude Edges

The next task is to model the walls of the excavation. You will select the edges of the hole and extrude them downward.

1. Switch to the Modify panel

2. Apply an Edit Mesh modifier

3. Switch to Sub-Object Edge. The edges of the hole will become selected

4. Set the User view back to a Left view, set the display mode to wireframe, then Zoom Extents and Region Zoom tight to the selected edges

5. In the Edit Mesh parameters, choose the button labeled Extrude

6. In the Extrude spinner, type -300 (this spinner uses inches as the unit, but you don't need to type any units), and hit Enter on the keyboard

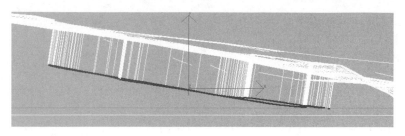

In the Left view you see that the edges of the hole haven't extruded straight down, but at an angle. To get the edges to extrude straight down, don't use the Extrude button, just drag the edges downward with the Shift key on the keyboard held down.

7. Use the Undo button to reverse the extrusion

8. Turn off the Extrude button

9. In the Left view, with the Move tool active, hold the Shift key on the keyboard as you drag the selected edges straight downward to a distance well below the surface of the terrain, as shown here. Be patient- the movement will be very slow

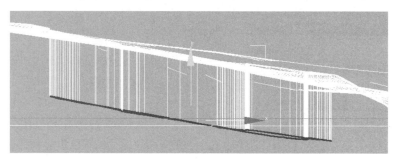

10. Switch to Sub-Object Vertex. The vertices at the top and at the bottom of the newly extruded faces will be selected. You want only the bottom vertices selected

11. With the Alt key on the keyboard held, drag a selection window (or two) that will deselect the upper vertices, as shown here:

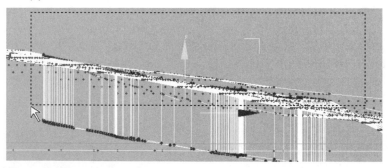

246

12. Click-hold on the Scale tool, and from the
    flyout choose Select and Non-uniform Scale

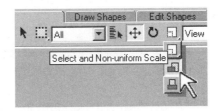

13. Position the cursor over the Y axis of the Transform Gizmo, and drag downward
    strongly. The vertices will begin to flatten out. Drag a few times, if necessary, until
    the vertices appear completely planar and horizontal, as shown here:

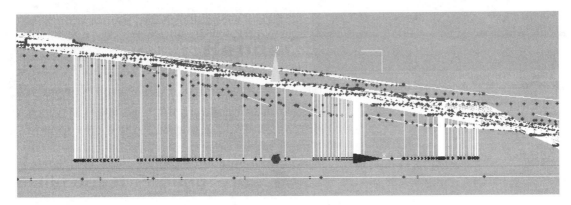

14. Choose the Move tool, and move the vertices upward until the right-side vertices
    meet the slope of the hill. Verify the correct elevation of the vertices by zooming in
    on the front of the terrace in the Perspective view

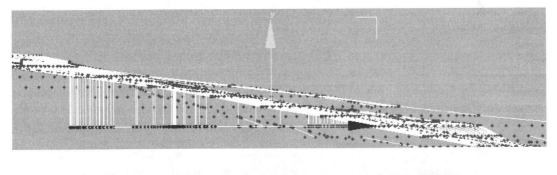

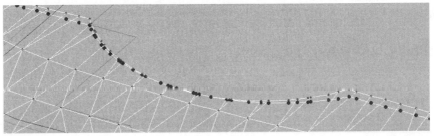

15. Turn off Sub-Object

16. Collapse the Stack (steps 3 through 6 on page 241) to reduce the terrain to an Editable Mesh

17. Navigate the Perspective view to see the area around the foundation hole

18. Choose the leftmost Render teapot. In the Render Design dialog box, choose the 800 x 600 preset render size button, uncheck the Save File check box, and choose Render

In the rendering you will see that the smoothing is not working correctly around the hole.

19. Close the rendered view (the Virtual Frame Buffer) and close the Render Design dialog box

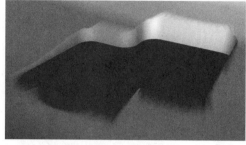

Smooth Modifier

20. Apply a Smooth modifier

21. Check the Auto Smooth check box

22. Choose Render Last - the rightmost teapot

The smoothing should be much improved.

23. Close the Virtual Frame Buffer

Terrace Floor

The floor of the terrace is easy - just apply a UVW Map modifier to the Slab shape, and move it up to the correct elevation.

1. Select the spline called SLAB.01

2. Apply a UVW Map modifier to make it solid

3. With the Perspective view active and set to shaded display mode, use the Z-axis of the Transform Gizmo to move the slab straight up until it is at the correct elevation. Zoom in tight and fine-tune the elevation by eye

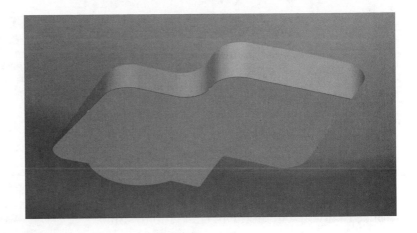

This completes the modeling of the terrain. You still need to model the pond, but this is a good point to switch gears for a while and work on natural materials.

Merge Lights

At this point in the process of developing a landscape, you would need to begin building a lighting scheme before starting on materials. If you develop materials with the default lights active, you will probably find that when you finally add lights, the materials will be wrong, and require extensive changes.

To move things along and keep you focused on landscape-specific subjects, a basic lighting setup has been created and saved into a .max file. Merge the lights into the scene.

1. Make the Top view active, Zoom Extents

2. From the menus, choose File / Merge

3. In the Merge File dialog box, browse to C:\VIZtutorials\Chapter5, highlight LightsChap5.max, choose Open

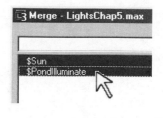

4. In the Merge dialog box, drag across both lights in the list to highlight them. Choose OK

5. Incremental Save the scene

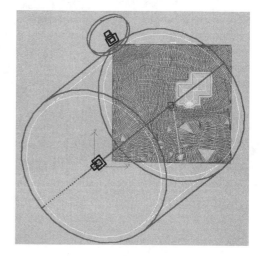

Two lights merge into the scene. One shines from the southwest, providing general illumination to the landscape, the roof planes of the house that you will merge in later, and to some extent, the walls of the house. This light casts shadows. The other shines from the northwest, and it has been set to only include the pond in its illumination, and it does not cast shadows.

Later you will add two more lights to the scene.

Landscape Multi / Sub-Object Material

The natural materials of a landscape offer an opportunity to explore some of the more advanced techniques in the Material Editor. At this point, many basic concepts that you will need to understand to make good materials are going to be put on hold, to be fully explained in the chapter on materials. In other words, you are about to be thrown into the deep end of the Material Editor. If some concepts elude you in this next section, don't let it bother you. Just follow the steps, getting a general feel for the process, and then maybe return to this section after you complete the chapter on materials. Before beginning this section, look at the glossary found after the last chapter (just before the index) and make sure you understand each of the terms presented in the glossary.

The material you will build will be a Multi / Sub-Object material containing three sub-materials for grass, road surface, and a forest floor material for the areas planted with trees. There are two approaches for mapping natural materials: use procedural maps or use bitmaps. Procedural maps do not employ a scanned image- they generate patterns entirely mathematically. Their strengths are that they do not get grainy when you zoom in on them (and bitmaps do), they do not show seams when repeated across a surface (bitmaps usually do), and there is no bitmap to keep track of (and possibly misplace or delete). Procedural maps are good for things at a distance, and bitmaps usually look more realistic for things up close. You'll try procedural maps first and then bitmaps in making the grass for this landscape.

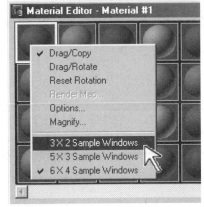

1. Open the Material Editor

2. You will only need six sample windows showing for this exercise. If only six are showing, you're set. If more than six are showing, right-click over the active sample, and from the menu choose 3 x 2 Sample Windows

3. With the first (the upper-left) sample window active, choose the Type button, which currently reads Standard, and from the Material/Map Browser, choose material type Multi / Sub-Object, then choose OK

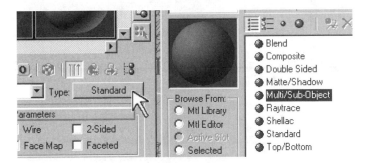

4. In the Replace Material dialog box, choose Keep old material as submaterial (not that it matters here, as you haven't worked on the old material), then choose OK

5. Choose the Set Number button, and set the number of submaterials to 3

6. Name the top level of the material Landscape Multi

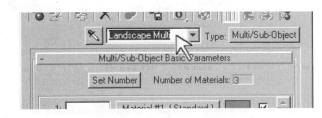

251

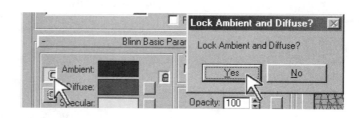

7. Name the three submaterials Grass, Road, and ForestFlr

Grass Material – Procedural

1. Choose the wide button next to the Grass name to drop down to that submaterial's parameters

2. Choose the link icon to the left of the Ambient and Diffuse color swatches, and in the dialog box, choose Yes to lock the two colors together. Then do the same for the Diffuse and Specular colors

3. Choose any of the three color swatches, and set the base color for this material to H = 90, S = 100, V = 75

4. Leave all other Basic Parameters at their defaults, and open the Maps rollout

You will use three maps channels to give the grass a natural appearance. You'll apply a large-scale Noise pattern to the Specular Level channel, so the land will be randomly brighter and darker. You'll copy that Noise into the Bump channel so that the land does not look impossibly smooth. Then you will use a complex map in the Diffuse Color channel to add layered patterns of different hues, at different sizes and textures, to suggest the natural variances of plant and earth color, grass density, and such in a field of grass.

Specular Level Map

A Specular Level map uses the brightness (value) of pixels in a map to determine where an object is brighter (areas of light pixels in the map), and where it is darker (areas of dark pixels). Without a Specular Level map, the bright areas of an object will be determined entirely by how you place lights in the scene. This map is an added level of control. A typical use of a Specular Level map is to make a wood floor appear worn (duller, darker) in places. Such a map (which you would make in a paint program) would be a light background with some dark smudges and streaks where the floor is worn or scuffed.

5. Choose the wide button labeled None in the Specular Level channel, and in the Material / Map browser, choose Noise, choose OK

6. You are now down one level, in the Specular parameters. Name this level Specular-Noise

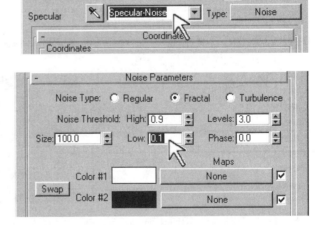

7. Set Noise Type to Fractal

   Set the Size spinner to 100

   Set the Noise Threshold High spinner to .9, and the Low spinner to .1

The Size spinner is in scene units; in this case inches. So the fractal noise pattern will begin a new iteration every 100 inches. The Threshold spinners allow you to do two things: control the definition at the boundary of the two colors, and weight the noise pattern toward either color. The closer the two spinners are in value, the more clearly defined the noise. If you decrease the High spinner more than you increase the Low spinner, Color #2 will become more dominant in the Noise pattern. If you increase the Low spinner more than you decrease the High, Color #1 becomes dominant.

253

8. Choose the Go to Parent button to return to the top level of the Grass submaterial

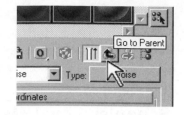

9. Drag the sample sphere for the Landscape Multi material onto the terrain in any view to assign the material to the terrain

10. Close the Material Editor, and Incremental Save

Camera

1. In the Top view, drag out a Target Camera, at any position

2. In the Modify panel, choose the 24 mm Stock Lenses button

3. With the Camera selected, make the Move tool active and right-click over it to open the Transform Type-in.

4. In the Absolute:World spinners, enter X = 350′, Y = 90′, Z = 10′ Keep the Transform Type-In open

5. Select the camera's Target

6. In the Transform Type-In, Absolute:World spinners, enter X = 320′, Y = 300′, Z = 10′

7. Close the Transform Type-In

8. Make the Perspective view active and type C to change it to a Camera view

9. Render the Camera view

254

You should see the effect of the Noise procedural map in the Specular Level channel. The effect is more pronounced in the distance than in the foreground. Cloning the Noise map into the Bump channel will give a more varied, natural appearance.

Bump Map

1. Open the Material Editor

2. Drag the Noise map from the Specular Level channel and drop it onto the Bump channel to clone the map

The cursor can be a bit confusing during this action. It shows a rectangle representing a map. Don't position the rectangle over the Bump channel. Instead, position the tip of the cursor's arrow over the Bump channel, and release.

3. In the Copy / Instance Map dialog box, choose Instance, so that if you change settings for the map in one channel, it will also change in the other

4. Change the amount (strength) at which the Bump map is used to 10

You are about to render the scene again to see the effect of the Bump map. As you develop materials, you may want to toggle between a before and after image to see the effect in the rendered scene of the changes you make in the Material Editor. You will set up the RAM Player to do this.

RAM Player

The RAM player is an image viewer with two channels, laid one over the other. You can load still images or movies into each channel, and compare the two channels by hiding and revealing the top one.

 RAM Player is a useful tool, but it is also one that tends to cause crashes, particularly in Windows 98. Save your scene before using RAM Player

1. To open the RAM Player, choose, from the menus, Rendering / RAM Player

2. Choose the teapot in the Channel A controls to load the last rendered image into channel A. In the RAM Player Configuration dialog box, leave all settings at the defaults, and choose OK

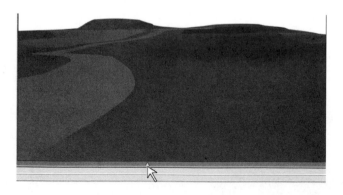

You can now minimize the RAM Player if you want, but don't close it.

3. Render the Camera view again

The terrain is noticeably bumpier. Compare the before and after.

4. In the RAM Player, choose the teapot in the Channel B controls to load the last rendering into channel B

5. Just above and just below the images are two small white triangles. Drag either triangle left and right to hide and reveal the Channel B image

6. Close the Ram Player

Diffuse Color Map: Mix

The Diffuse area of an object is the area of general illumination.

The shiny spots are the Specular Highlights

The part in shadow is called the Ambient area.

A Diffuse Color map just determines the basic qualities of a surface- a pattern of colors. In the case of this grass, the Specular Level map and the Bump map have done a good job of adding natural variance in brightness to the land and making it appear to undulate, but the land is all one hue. A Diffuse Color map made by blending some shades of light and dark greens and browns will make the grass more believable.

1. In the Material Editor, in the Maps rollout of the Grass sub-material, choose the bar labeled None in the Diffuse Color channel

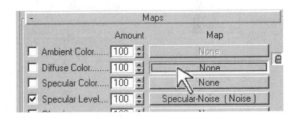

2. In the Material / Map Browser, choose Mix, then choose OK

The Mix map does what the name implies- it mixes two colors or two other maps together over the surface of the object. The Mix Amount spinner determines what percent of the second color or map is mixed into the first. If the mixing is done just by the Mix Amount spinner, the blending occurs evenly over the whole surface. More commonly, a grayscale map is used to do the mixing. Wherever the pixels of the mix map are darker, the first color or map dominates, and wherever the pixels of the mix map are lighter, the second color or map begins to show more strongly.

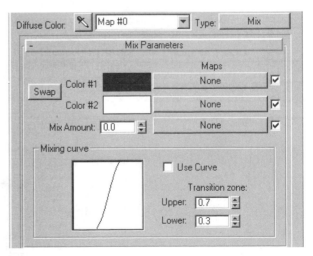

For this grass material, you will mix together a Smoke map and a Dent map, and a Noise map will be used to do the mixing.

Smoke Map

3. Choose the button labeled None next to the Color #1 swatch

4. From the Material / Map Browser, choose Smoke, then choose OK

5. You are now down one more level in the Material Editor. Name this level Earthtones

6. Leave all parameters set at their defaults. Choose the color swatch for Color #1

7. Set Color #1 to:
    H = 40
    S = 90
    V = 90

8. Leave the Color Selector open, choose the Color #2 swatch

9. Set Color #2 to:
    H = 45
    S = 120
    V = 60,
    then close the color selector

10. Choose the Go to Parent button

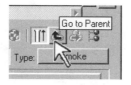

Dent Map

11. Back up at the Diffuse Color level, choose the bar labeled None next to the Color #2 swatch

12. In the Material / Map Browser, choose Dent, choose OK

13. Name this level of the Material Editor Greens

14. In the Dent parameters, set the Size to 100, the Strength to 10, and the Iterations to 1

The Size spinner sets the size of the dents. The Strength spinner alters the number of dents; the higher the Strength, the more dents per unit area. Iterations controls the complexity of the dents. Higher Iteration numbers create dents in the dents. Higher numbers also take longer to calculate. There aren't really any reliable rules of thumb for how to set these three settings – it is purely trial and error.

15. Choose the Color #1 swatch, and set the color to:
    H = 90
    S = 140
    V = 80

16. Choose the Color #2 swatch, and set the color to:
    H = 85
    S = 40
    V = 130

17. Close the Color Selector, and choose Go to Parent

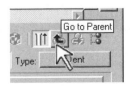

Mixer Map

You have set up the two procedural maps to be mixed, now all that is left is to describe the map that will do the mixing. Without this mixer map, the Earthtones Smoke map and the Greens Dent map would blend together uniformly over the terrain, and the result would be muddy. The mixer map will be Noise. Where the noise pattern is darker, the Earthtones Smoke map will show more strongly, and where the noise pattern is lighter, the Greens Dent map will dominate.

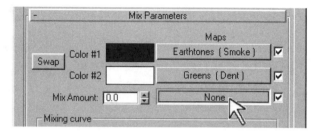

1. Back up at the Diffuse Color level, choose the bar labeled None in the Mix Amount channel

If you did not want the two maps mixed in random patches, and simply wanted the two maps blended together uniformly, then you would just use the Mix Amount spinner, which uses a zero-to-100 scale. Set to 50, the two maps (or colors) would blend with equal intensity.

2. In the Material / Map Browser, choose Noise, then choose OK

3. In the Noise Parameters, set Size to 150. Leave other parameters at their defaults

The earthtones and the greens will be mixed in soft-edged patches that begin a new random iteration every 150 inches.

4. Choose the Go to Parent button

5. Incremental Save

6. Render the camera view

260

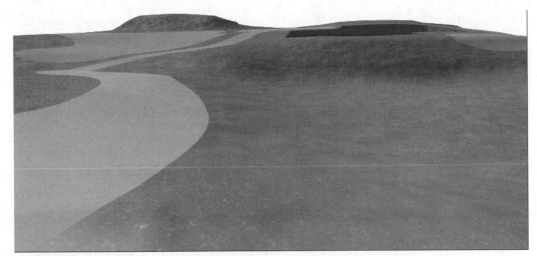

While this grass does not look like a photograph of grass, it does look convincingly natural, and in the distance it does approach a realistic look. Up close you would expect to see a carpet of individual blades of grass, and procedural maps cannot do that. If photorealism is what you want, you'll need to make natural materials based on photos.

Bitmapped Materials

The trick to making realistic grass using bitmaps is to start with the right bitmap. Well-made, high-resolution bitmaps of grass are hard to find, and the best thing to do is to make them yourself, which of course requires a knowledge of a paint program. For source photos of grass, nothing beats a trip to a golf course. Take a digital camera, set it at high quality and resolution, and take shots of uniform areas of the fairway, taking a few shots of each spot, from different angles.

In a paint program, double the canvas width, then mirror the bitmap horizontally, so that when the image repeats, there will be matching pixels at the seam.

Double the canvas height, and mirror the two tiles vertically, so that the bitmap will have matching pixels at the seams in both directions

Using the cloning tool (rubber-stamp), clone grass to hide the two interior seams. Stay away from the edges when you do this, or the outside edges won't match when the bitmap is repeated. The areas of grass you clone from can come from this image, but it sometimes works better to clone from another photo with similar grass.

Once you've mirrored in both directions and disguised the inside seams, you need to test the bitmap for proper tiling. Mirror the entire new image, in each direction, and see how noticeable the seams at the edges are. If there are distinct objects, like leaf clutter or weeds, at the very edges, they will be noticeable at the seam, and you need to edit these out. With patience, you should be able to make small edits at the edges of the bitmap so that when it tiles, the seams are not immediately noticeable.

The final product of this effort is a high-resolution bitmap that repeats, or tiles, reasonably well. A real pro might succeed in making the seams perfectly hidden, but it's not really necessary. If you get two images that tile fairly well, and you make a Mix map mixing those two images using Noise, as you've just done with the procedural maps, the seams will be indistinguishable and the grass will be realistic.

There are maps prepared as described above on the CD that accompanies this book, and you will use two of them in this material. Rather than build the entire material from scratch, copy the procedural grass and modify it into a bitmapped material.

1. In the Material Editor, choose Go to Parent until you are at the topmost level of the Landscape Multi material

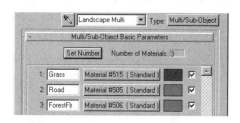

2. Drag the material's sample sphere and drop it into the next available window to make a copy of the material

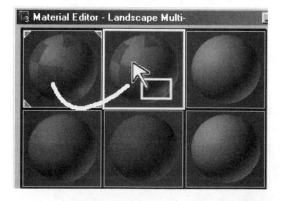

Notice the difference between the sample window of the original, and the sample window of the copy. The original has white triangles at the corners, which tells you that

262

the material is "hot"– it is applied to an object in the scene, and any changes you make to that material will update immediately in the scene. The new copy has no triangles– it is "cool", and you are free to make changes without affecting the scene.

3. Name the material Landscape Multi-2

4. Drop to the first submaterial, Grass

5. Drop to the Diffuse Color level, the Mix map

6. Drop to the Color #1 map, the Earthtones Smoke map

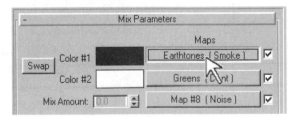

7. Choose the Type button, labeled Smoke, and from the Material / Map Browser, choose Bitmap, then choose OK

8. In the Select Bitmap Image File dialog box, browse to C:\Viztutorials\Chapter5, highlight Grass_Clover_01.jpg, and choose Open

The Grass_Clover_01.jpg bitmap, at pixel dimensions of 2304 x 1500, is a very large bitmap- certainly many times larger than what is needed for the renderings you are doing in this tutorial. It has been left so large in order that you can use it in projects in which you will be rendering at very high resolutions, perhaps for poster-size prints. The larger the output size you choose in the Render Design dialog box, the larger the bitmaps used in materials need to be. If you plan to use this bitmap in projects that are to be rendered at lower resolutions, you should probably resize the map in a paint program and save a lower-resolution version under a different name.

9. Choose the Go Forward to Sibling button to move to the Color #2; Dent level

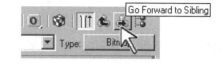

10. Choose the Type button, labeled Dent, and from the Material / Map Browser, choose Bitmap, then choose OK

11. In the Select Bitmap Image File dialog box, browse to C:\Viztutorials\Chapter5, highlight Grass_Clover_07.jpg, and choose Open

12. In the Coordinates parameters for the bitmap, set the Angle / W: spinner to 90

Rotating the direction of one of the two bitmaps will help make the tiling less obvious.

13. Choose Go to Parent twice to return to the maps channels of the Grass submaterial. In the Maps channels, at the left of the Bump channel, uncheck the check box, to deactivate the Bump map

Since the bitmap of grass already shows the natural bumpiness of grass, adding additional bumpiness with a map is too much.

That's all you need to do for this new material (except assign it to the terrain). The two bitmaps will be mixed together by the same Noise map that was used to mix the procedural maps, and the Noise map used in the Specular Level channel will also work well for this bitmapped material.

13. Drag the new landscape material from the sample sphere and drop it onto the terrain in any view to assign it

The procedural-mapped version of the grass used no bitmaps, therefore it needed no UVW Map modifier. This is another advantage of procedural maps: you can control their placement with a UVW Map modifier if you want, but it is unnecessary, and it usually works best to let the object's own local coordinates (Object XYZ) control the map. Now that there are bitmaps involved in the new grass material, you will need a UVW Map modifier applied to the terrain.

14. In the Modify panel, with the Terrain selected, apply a UVW Map modifier. Leave the mapping type set to Planar, and set the Length and Width to 50 feet

15. Render the Camera view

Now the grass is fairly believable; there are almost no visible seams and there is the variance that is needed to make it look natural.

While the Mix map setup might be a bit confusing the first time you try it, it is a tool you need to understand in order to make natural-looking materials. In these grass materials you used a simple Noise map to mix the other two maps together. In many cases you will need a map that controls the mixing in a more specific way, defining distinct areas of one map that mix into areas of another, and you need a decent knowledge of a paint program to create such a bitmap.

You could use the same principles used in the grass submaterial to make materials for the road and the forest floor, but in order to move things along, just pull a couple pre-made materials from a library to finish the Landscape Multi material.

Populating Submaterials

1. In the Material Editor, choose Go to Parent until you are at the top level of the Landscape Multi-2 material

2. Choose the Get Material button

   ![Get Material button]

3. Position the Material / Map Browser next to the Material Editor

4. In the browser, choose Browse From: Mtl Library

   ![Browse From options: Mtl Library, Mtl Editor, Active Slot, Selected, Design, New]

5. Choose File / Open

6. Open Ground.mat

7. If the Browser shows red and green parallelogram icons as well as blue sphere icons, check the Root Only check box

8. From the Ground.mat library, drag Ground-Grey Dirt 3 and drop it onto the bar for submaterial #2, the Road material

9. Drag Ground-Mulch 1 and drop it onto the bar for submaterial #3, the Forest Floor

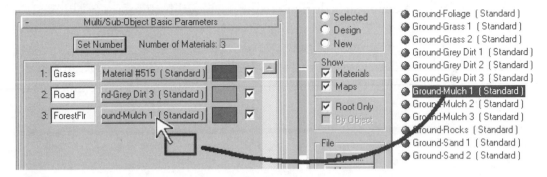

10. Close the Material / Map browser and the Material Editor, save the scene, and render the Camera view

The obvious problem at this point is the razor-line border where grass meets road and forest floor. Admittedly, there is not an easy fix for this. The least complicated way is to simply blend those parts of the final rendered image in a paint program, if what you're producing is a still image. What is needed in VIZ is functionality at the top level of the Multi/Sub-Object material that allows you to define some blurring/blending between areas of faces that carry different material IDs. Hopefully such a feature will appear in a future release.

To blur material boundaries, it is possible to render a version of the landscape from the

Top view, in which areas of grass are black, roads are white, mulch areas are black, and so on, then apply some blurring to this image in a paint program, and use the image as the blending bitmap in a complex "nested Blend" material (a Multi / Sub-Object material in which each submaterial is a Blend type). The procedure is hard to describe and envision, and harder to carry out because it is difficult to get the blending bitmaps positioned just right.

To hide hard-edged boundaries, try to position objects to obscure borders. Put some rocks, grasses, and trees in the way, and blur any obvious boundaries in a paint program.

11. Close the rendered view

Environment Map

There is an image in the \Chapter5 folder of evergreen-covered hills and blue sky that will fit this scene well.

1. From the menus, choose Rendering / Environment

2. Choose the Environment Map bar, labeled None

3. In the Material / Map Browser, choose Bitmap, then choose OK

4. Browse to the C:\VizTutorials\Chapter5 folder, highlight ForestMtns.jpg, and choose Open

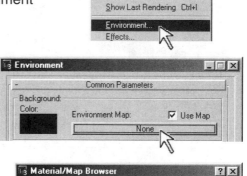

268

5. Close the Environment dialog box

6. To see the Environment Map displayed in the Camera view, make the Camera view active, then from the menus choose View / Viewport Image / Select

7. In the Viewport Image dialog box, check Use Environment Background and Display Background, then choose OK

8. Render the Camera view to see the background

Suppose you wanted to move the background image up a bit, or flip it horizontally. To do so, you would keep the Environment dialog box open, open the Material Editor beside it, and drag the name of the Environment Map from the Environment dialog box to an unused sample window in the Material Editor. In the Material Editor you could then use the Offset spinners to nudge the image around. U means the X direction, V means the Y direction. So entering .1 in the V: spinner moves the background image up 10%.

Tiling is how many times the image repeats, and it is how you control the scale of the image relative to the view. If you wanted the width of the background enlarged twice, the U: Tiling spinner would be set to .5. To make the image fill two-thirds the width of the view, the U: Tiling spinner would be set to 1.5 (because 1 divided by .666 = 1.5). That is a bit counterintuitive at first. Just remember that smaller Tiling values scale an image up, and larger Tiling values scale an image down. If you were to set both Tiling spinners to 1.5 so that the background image becomes smaller, you would probably want to uncheck the two Tile check boxes, because you probably would not want the image to repeat (tile), but just to fill less of the view.

In still images, you generally want the background image to just sit behind the geometry of the scene, projected perpendicular to the camera's line of sight. This is called Screen mapping the environment. But in an animation, or in a Smoothmove panoramic rendering (a 360-degree interactive view), you certainly wouldn't want an unchanging background behind the geometry as you look about. In the Mapping drop-down list are choices Spherical, Cylindrical, Shrink-wrap, and Screen, with Screen being the default. Spherical wraps the image onto an unseen sphere around the geometry, and Cylindrical wraps it onto an unseen cylinder (appropriate if you don't plan to look up). Both of these types will leave one seam in the environment, where the edges of the the environment map meet. You can move that seam behind the camera with the U: Offset spinner. Shrink-wrap is like a balloon; the map is wrapped down over a sphere, with a puckered gather at the bottom. Since that gather is below the ground, and therefore unseen, Shrink-wrap may be a good choice for an interactive view. However, Shrink-wrap also results in the most distortion of the environment image – imagine having to stretch a sheet of thin rubber to wrap it balloon-fashion around a basketball.

Water

The surface of a body of water is easy to depict, and water always adds drama, realism, and a sense of life and movement to an image. There are two approaches to modeling the surface of water. If you need big waves and swells, or you have only small waves but the edge of the water will be clearly visible (as in a swimming pool), then you'll have to build some waviness into the geometry of the surface. More often in a design visualization, the water is fairly calm, or the water's edge is far away and it won't be noticed if the edges don't show ripples or small waves. In that case, water is achieved entirely in the material; the geometry can be a flat plane. You'll use the flat plane method for the pond in this scene.

1. Type H, select POND.01

2. In the Modify panel, apply a UVW Map modifier to turn the pond shape into a mesh, and to position any bitmaps that might be used in the material

3. From the menus, choose Modify / Properties

4. In the Object Properties dialog box, in the Rendering Control group, choose the button labeled By Layer to change it to read By Object, and to make the various settings available

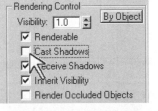

5. Uncheck Cast Shadows, choose OK

6. Open the Material Editor

7. Make the third sample window active

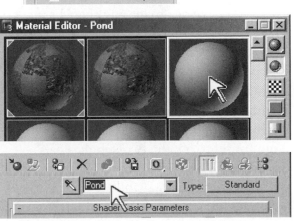

8. Name the material Pond

This will be a fairly complicated material. First, it will be a Blend material, because there needs to be a blending of a water material into a sandy shoreline material (you can't have the water just end in a sharp edge at the grass). This is a very specific blending task, a narrow blended band that follows the shape of the pond's edge, so it will be accomplished with a bitmap showing the pond from above, with the pond rendered totally black (which calls up the water material) except at the edges, where it blends into totally white (which calls up the sand material). This blend map will take care of the water-to-sand blend, but what about the sand-to-grass transition? This wouldn't look right as a sharp edge either. The sand material will incorporate a special bitmap also- an Opacity map. This is another rendering from above, with the pond shown white (opaque) except at the very edge, where it quickly fades to black (transparent). This will cause the sandy shoreline to feather into the grass in a narrow band.

If this is not all crystal-clear, don't worry; the explanation will be expanded as you work through the steps. Then come back and read it again, and it should be much clearer.

Blend Type

1. Set the material Type to Blend

In the Replace Material dialog box, it doesn't matter whether you choose to keep or discard the old material, because you have not made any changes to the old material.

The Blend Material blends two submaterials. If you just want a general blending of the two submaterials, use the Mix Amount spinner. With Mix Amount set to 50, the two submaterials meld equally and uniformly (think of putting chocolate and vanilla pudding in a blender until you have tan pudding).

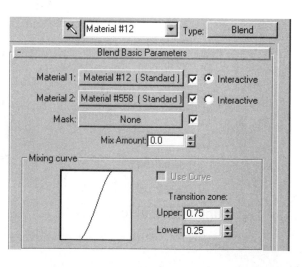

272

If you instead assign a map in the Mask channel, you control where the first material dominates, and where the second one does. Think of dropping a spoon of vanilla pudding into chocolate, then swirling it into a design. For a very natural blending pattern the Mask map might be Noise. In the case of the water blending at its edges into the sandy shore, you will use a bitmap from the \Chapter5 folder that looks like this:

The black calls up the water material and the white calls up the sandy shore material. The gradient at the edge blends the two materials.

When a bitmap is used to do the blending, the Mixing Curve allows you to alter the values (brightness) of the bitmap without needing to edit the map in a paint program. The map will not be changed; the Mixing Curve just applies multipliers to the values of the map to change its effect as a blending map.

To the right of the Material 1 and Material 2 channels are buttons labeled Interactive. The blending of the two materials will not be shown in shaded views (the Interactive viewports). Whichever submaterial has the Interactive button chosen will show. To see the blending effect, you need to render.

First Submaterial: Water

1. Select the button in the Material 1 channel to drop to that submaterial level, and rename that level Water

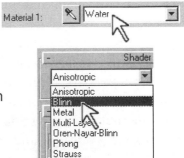

2. Open the Shader Type drop-down list and set it to Blinn

Shaders will be fully explored in the chapter on materials.

3. In the Shader Basic Parameters, lock the Ambient and Diffuse color swatches together, if they are not already locked

4. Select either color swatch and set the color to:
   H = 160, S = 200, V = 25

This is a very deep, dark blue. You will add reflection to the water, and reflections are usually more vivid when the base color of the object is dark. In water, the Ambient and Diffuse colors will be the color of the troughs of waves.

5. Set the Specular color swatch to:
   H = 0, S = 0, V = 255

The Specular color will be the color of the crests of waves.

6. Set the Glossiness spinner to 15

Glossiness is the diameter of the specular highlight. For a broad, flat surface like a pond or a floor or wall, you generally want the Glossiness set low.

7. Open the Maps rollout

You'll use three of the channels in making the water. For the appearance of small wavelets, you will need a Bump map, which will be Noise. You will clone this Noise into the Specular Level channel, so that the waves will shine brightly at their crests, and be dark at the troughs. And you'll assign a Raytrace map in the Reflection channel. It is reflection that really makes water appealing and convincing.

Specular Level Map: Noise

8. Choose the wide button in the Specular Level channel

9. In the Material / Map Browser, choose Noise, then choose OK

10. In the Noise Parameters, set the Noise Type to Fractal and the Size to 40. Leave everything else as is

11. Choose Go to Parent

Bump Map: Noise

12. Drag the Noise map from the Specular Level channel and drop it over the Bump channel

13. In the Clone (Instance) Map dialog box, choose Copy, choose OK

In many cases Instance is the best choice here, so that if you make changes to the map parameters in one channel, they will update automatically in the other. But in this case, you need to make the Bump map slightly different from the Specular Level map, so Copy is the correct choice.

14. Set the spinner for the value at which the Bump map is used to 20

15. Drop to the Bump level

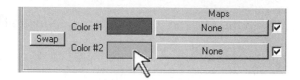

16. In the Noise Parameters, select the
color swatch for Color #1, set it to:
    H = 0, S = 0, V = 90

    Set Color #2 to:
    H = 0, S = 0, V = 175

Making these color values closer to each other will make the Bump map less intense.
You could instead simply lower the value at which the Bump map is used, maybe setting
it as low as 2 or 3, but changing the color
values of the map offers more control.

17. Select Go to Parent

All that's left now is to add some reflection.

Reflection Map: Raytrace

18. In the Maps channels, set the spinner
for reflection amount to 30

19. Choose the button in the Reflection
channel, and in the Material / Map
Browser, choose Raytrace; choose OK

20. For now, leave all the Raytrace
parameters as they are, and
choose Go to Parent twice, to
return to the top level of the
Blend material

Second Submaterial: Shoreline

This submaterial is easy– there is a premade sandy shore material in a library that will work perfectly.

1. Select the Get Material button

2. In the Material / Map Browser, choose Browse From: Mtl Library

3. Choose File / Open

4. In the Open Material Library dialog box, highlight 3dsviz.mat. Choose Open

5. If it's not checked, check Root Only

You don't need to see a listing of every bitmap incorporated into every material.

6. Drag material Ground-Sand Tex. from the library to the button for Material 2

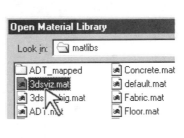

7. Save the scene

The next step is to assign the bitmap that will control how and where the water blends into the sandy shore.

Blending Map

1. In the Material / Map Browser, switch from
   Browse From: Mtl Library to Browse From: New

2. Drag the word Bitmap from the Material / Map Browser and drop it onto the button labeled Mask

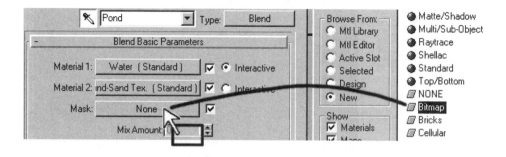

3. In the Select Bitmap Image File dialog box, highlight PondBlend.tga, then choose the button labeled View at the lower-left of the dialog box

This bitmap was made by making a totally black material, applying it to the pond, then making a totally white material, applying it to the terrain, then rendering in the Top view, at a resolution of 1024 x 768 (a moderately high resolution, so the bitmap is detailed enough to convey smooth gradients). The rendered bitmap was saved to the hard drive, opened in Photoshop, and the Gaussian Blur filter was applied at a value of 18, to create a very soft gradient all around the edge of the water. The image was then cropped to the outside edges of the gradient. The Blend material will use this map to blend the water into the sand along the gradient. The blending will make the water appear to get shallower as it reaches the shore.

4. Close the window showing PondBlend.tga

5. In the Select Bitmap Image File window, make sure PondBlend.tga is still high-lighted, and choose Open

⚠ The Mask: button should update to show the label PondBlend.tga, but it may not do so (particularly if you use Windows98). Don't worry, it's just a little glitch in the program. When you change to another area of the Material Editor and then return to this level, you'll see the button properly updated.

6. Close the Material / Map Browser

The material only needs a couple more changes; see how it looks so far.

7. Make sure POND.01 is still selected in the views, then choose Assign Material to Selection (or just drag the material onto the pond surface in a view)

8. Save the scene, then render

⚠ Now that there is a Raytrace map present, you will see a window appear as the renderer reaches the pond. Do not try to cancel the rendering when the Raytrace Engine Setup window is visible– VIZ might lock up, and you'll have to Ctrl-Alt-Del to shut VIZ down.

There are three problems to solve. The first is the knife-edge transition between sandy shore and grass. The second is that the pond is very dark. It will look much livelier once you add an accent light. The third problem is that the reflection is wrong. You are seeing the closest hill correctly reflected near the opposite shore, but you should also be seeing the blue sky reflected, and instead you are seeing evergreens on a hillside.

You are using a Raytrace map to create the reflection. There are two other maps that can be used to create reflections: the Flat Mirror map and the Reflect/Refract map. None of the three types of reflection calculates the reflection of an Environment map in a consistently correct and predictable way. Remember that the distant hills and blue sky are present in your scene as an Environment map. The error that the renderer often produces is a projection, rather than a reflection, of the environment onto the reflective surface. If you view the ForestMtns.jpg (which you can do via Tools/Display Image), you will realize that what is showing up on the surface of the water is the area of ForestMtns.jpg that you would see if you had x-ray vision and could look through the pond to what is behind it. It would certainly be welcome if this situation were corrected in a future release of VIZ.

For now, there are two ways to fix it. The method that will work for all types of reflection is to draw a large rectangle, apply a UVW Map modifier to it, stand it up behind your scene, and apply a material with the environment map in the Diffuse Color channel onto the rectangle screen. A big Hollywood backdrop, basically. It takes more work to get the image positioned just right than if you just assign it as an Environment map, but the fact that the image is sitting on some geometry will make reflections work correctly. If the reflection map you're using is Raytrace, as in this scene, there's a nice trick in the Raytrace parameters to get around this problem quickly and easily.

Correcting the Environment Reflection

1. Close the rendered view

2. In the Material Editor, drop into the Water material

3. In the Maps channels, drop to the Reflection level

4. In the Background group, choose the button next to the wide bar labeled None, then select the bar

5. In the Material / Map Browser, highlight Bitmap, and choose OK

6. Browse to C:\VIZtutorials\Chapter5, highlight ForestMtns.jpg, and choose Open

7. Open the drop-down list labeled Mapping and set it to Screen

8. Set the Tiling V: spinner to -1

The minus will flip the image upside-down.

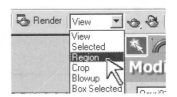

When you re-render to see the effect of these settings, why render the entire view? You only need to see the pond area updated. Render the pond area with Render Region.

Render Region

1. Between the three Render teapots, open the Render Type drop-down list and set it to Region

2. With the Camera view still active, choose the middle teapot, Quick Render. A dashed rectangle appears in the Camera view

3. To move the region, place the cursor anywhere inside it, so the cursor is a four-way arrow, and drag. To size it, drag on the squares at the corners or sides

4. Once the region surrounds the pond area, choose the OK button in the lower-right of the Camera view.

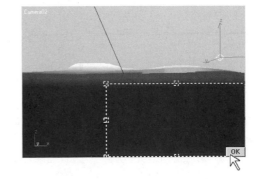

⚠ The OK button tends to disappear, especially in Windows98. If it does, just click in the very lower-right, where you know the button should be– it will work.

Another highly useful choice in the Render Type dropdown list is Selected– only the object selected in the views is rendered. Render Selected and Render Region are crucial tools for being productive.

The reflection looks correct now.

The last thing to do in the material is to have the sandy shore disappear at its very edge, so it blends into the grass and looks natural. This is done with an Opacity map.

Opacity Map

1. Close the rendered view

2. In the Material Editor, choose Go to Parent until you are at the top level of the Blend material

3. Drop to the Ground-Sand Tex. sub-material

4. Open the Maps channels and choose the Opacity channel button

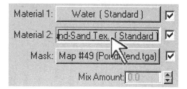

5. In the Material / Map Browser, choose Bitmap, and choose OK

6. In the Select Bitmap Image File dialog box, select PondEdge.tga, and choose Open

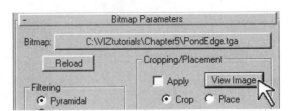

7. In the Bitmap Parameters, choose View Image

Like the PondBlend.tga, this image is a rendering from the Top view, with a white material on the pond, which will make the geometry opaque, and a black material on the terrain, for transparency. It was also blurred in Photoshop, using Gaussian Blur set to a value of 4, and was then cropped to the outside of the gradient. The gradient is thin and just barely visible in this figure. The gradient edge will make the pond edge dissolve and blur into the grass.

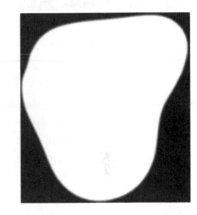

8. Close the window showing PondEdge.tga

9. Render Region the pond area. Region should be showing in the drop-down list from the last render, so just select Render Last

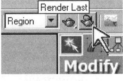
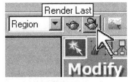

The pond is looking fairly good- everything is blended in a natural way.

If you detached the road as a separate object (using an Edit Mesh modifier), and scaled its width a bit so it overlapped the grass, you could use the same bitmap blending techniques to blend from road surface to shoulder to grass. But it would not be as easy as the pond.

Well, that's it for this rather complicated material. Natural materials for landscapes are some of the most challenging materials to make (and understand), and if you have been able to follow what's been happening the last thirty pages, then well done. You have gained a level of comfort with VIZ that should allow you to handle most challenges you will face in typical design visualizations. Materials will be further explored in the final chapter of the book, and it might be a comfort to you to know that what you've just done is as complex as anything in that chapter. If some of this chapter's exploration of natural materials overwhelmed you, don't sweat it; the concepts will come up again in the final chapter on man-made materials, and if you revisit this chapter after working through that one, you should find it much easier to follow.

There is one thing left to do to the pond to make it more interesting. It involves a light, not materials. Water surfaces look good with a streak of light coming toward the camera. You can control exactly where that streak sits with the Place Highlight tool.

Place Highlight

This an easy tool to use, and one of the most clever. You create a light, then point through the camera to the point on any surface you want that light to highlight, and the light moves to position itself at the correct angle to create that highlight.

1. Close the rendered view, close the Material Editor, and save the scene

2. Make the Top view active, Zoom Extents

3. In the Create panel, Lights category, choose Omni

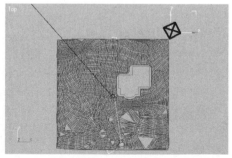

4. Click to place the Omni just off the north-east corner of the terrain

5. Switch to the Modify panel

6. Uncheck Cast Shadows

7. Set the light's HSV spinners to:
   H = 0, S = 0, V = 180

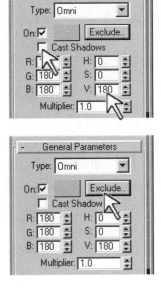

8. Choose the button labeled Exclude

An indispensible trait of VIZ lights is the control you have over what a light "sees," and what a light ignores. You do not want this light to illuminate anything but the pond.

9. In the Exclude / Include dialog box, highlight POND.01 in the left list and choose the arrow pointing right, between the two list boxes, to transfer POND.01 to the right list

10. Choose the Include button, then choose OK to close the dialog box

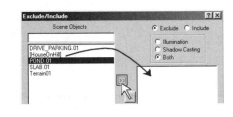

This light will now only illuminate the pond. You will work more with the Exclude/ Include features in the chapter on interior lighting.

11. Make the Camera view active

You are going to need to drag the cursor over the surface of the pond, and the terrain is in the way. If you Lock the terrain, it will be ineligible to be selected or picked on.

12. If the Display Floater is not already open (and probably parked in the very lower-left of the VIZ interface), from the menus choose Tools / Floaters / Display Floater

13. In the Display Floater, choose Lock / By Name, and in the Lock by Name dialog box, highlight Terrain01, and choose Lock

14. With the Omni light still selected, choose, from the vertical toolbar on the left, the Place Highlight tool

15. Place the cursor anywhere on the surface of the pond, and drag. A blue vector will move across the pond surface, and the Omni light will move as you drag, positioning itself to create a highlight wherever the vector is. Drag the vector to the very right edge of the Camera view, and near the shoreline in the foreground, and release to set the light

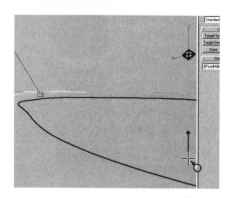

16. Save the scene

17. Render the Camera view (use the Render Last button)

There is now a gleaming stripe on the water, which makes it more convincing and interesting. Does it make sense to have this gleaming area on the water? No, because there is no brilliantly lit area of the sky above it. To make this look correct in this scene, you would need to open the background image in a paint program and add a bright part at the right of the sky. But even without editing the background, it is still worth adding this highlight, as it makes it a better picture, and most viewers are not going to scrutinize the image for logical accuracy. They just want to get a certain feeling from the image.

Trees

There was a brief description in chapter 2 (page 67) of options for populating a scene with trees. In short, these are the choices:

Use 3D, computer-generated trees. VIZ comes with a few of these, and there are several vendors selling 3D tree models.

Use a plug-in that generates foliage during rendering. In the views, the trees are represented by simple geometry. At rendering, all foliage is generated.

Use RealTrees, a plug-in that places a stand-in in the view, and at rendering retrieves one of 360 photos of a tree, depending on your camera's viewing angle.

Use photos of trees mapped on to simple planes – "cutouts".

Do not add trees in VIZ – add them in a paint program.

The first two options create computer-generated trees, and the last three options work with scanned photos of trees. Mixing the two methods in a scene is not advisable; go with either all computer-generated or all photographic. Two factors may steer you away from all computer-generated. First, they require large amounts of system resources, and they usually render very slowly. Second, clients tend to dislike them. This is of course a subjective observation, and you may have a client who thinks they look great. But the computer aesthetic has not totally permeated the design industries, and there are many traditionalists who recoil at imagery that is overtly computer-generated. If you want the client to be happy the first time, photographic trees are the best bet.

If you have decent skills in a paint program like Photoshop, and you are only rendering one or two still images of the scene, consider adding much (or all) of your foliage "in paint" (as they say in the business). Placing and orienting dozens or hundreds of cutout trees in a VIZ scene can be very time-consuming, and the rendering process will slow considerably when all the trees and their shadows need to be calculated into the image. It is often easier to just paste all the foliage (and its shadowing) in a paint program. The ultimate paint program for adding foliage and entourage of all kinds might be Piranesi (www.informatix.co.uk). With Piranesi, you drag entourage from an extensive library, and when you drop it into the image, it scales itself correctly. Move the tree or person in the image, and the scale updates to fit the perspective. Move a tree partially behind another object, and it will be properly obscured by that object. Addition of shadows in Piranesi is almost automatic. You should investigate this remarkable program.

Cutout Trees

The most common method for placing trees in the VIZ scene is to use cutouts. In this method, the geometry is a simple plane. A picture is applied to that plane in two ways: once to display a photo of a tree (diffuse color map), and once to determine where the plane should be seen, and where it should be invisible (opacity map).

Cutout Shape

There are two choices for the shape of the cutout plane: rectangular or tree-shaped. Which you use depends on what sort of shadow you want the tree to cast. Remember from Chapter 2, pages 81-82, that of the two types of shadows, Shadow Map and Ray Traced, only ray traced shadows handle transparency correctly. There is transparency at work in a cutout; the opacity map is making the plane disappear where there is no tree. If the geometry of the cutout is just a rectangle, a light casting shadow maps will result in a rectangular shadow on the ground, because the shadow map will not respect the opacity map of the cutout. To get a tree-shaped shadow when the cutout plane is rectangular, you must use ray traced shadows. Ray traced shadows are crisp-edged, which is fine if the scene shows a bright, sunny day. If the day is partly cloudy or at twilight, you need softer shadows, and that means shadow maps. To use shadow maps with cutouts, the plane of the cutout needs to have the general shape of the cutout item.

Single Versus Double Cutout Planes

Cutouts only look good viewed from a nearly perpendicular point of view. If you get too far around one side of the cutout, you can see that it is just a flat plane. There are two solutions to this. One is to assign instructions to each cutout telling it to rotate around to always stay perpendicular to the camera. These instructions are called a Look At controller, and setting up a Look At controller, while not particularly difficult, is beyond the scope of this book. See the online reference for instructions on the Look At controller. The more common solution is to use two planes in a cross for each cutout, so that any viewing angle shows a tree shape. At certain angles you get an on-edge view of some planes, but they usually blend into the foliage of the cutout and are not noticeable. These cross-plane cutouts work quite well at a distance.

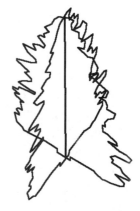

Scatter

You will use a creation tool called Scatter to distribute cutout evergreen trees in the two tree areas. Scatter will use a copy of the terrain as a Distribution Object, and you'll set Scatter to only use the faces of the tree areas when scattering trees. Scatter allows you to randomly alter the scale in any direction and the rotation of each tree, making it very easy to quickly get a natural-looking forest.

1. In the Display Floater, choose Unlock / All to unlock the terrain

2. Make the Top view active

3. From the menus, choose File / Merge

4. Browse to C:\Viztutorials\Chapter5, highlight FirTree01.max; choose Open

5. In the Merge-FirTree01.max dialog box, highlight FirTree01; choose OK

The double cutout evergreen shape shown on page 288 is now merged into the scene.

6. Select Terrain01

7. In the Modify panel, apply a Mesh Select modifier

8. Choose sub-object Polygon

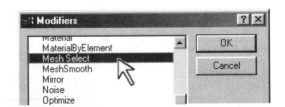

9. Set the Select by Material ID spinner to 3, then choose the Select button

The faces of the two tree areas should be selected.

10. Type H, select FirTree01

Note that you did **not** leave sub-object mode before typing H to select the tree. The terrain is still in a state of sub-object selection, which is what will allow Scatter to use only the tree area faces as the scatter area.

11. In the Create panel, geometry category, open the sub-category drop-down list and choose Compound Objects

12. Choose Scatter

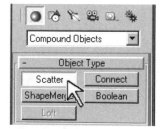

13. Choose the Pick Distribution Object button

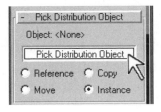

14. In any view, pick on the terrain

15. Switch to the Modify panel

16. To quickly scroll to the bottom of the Modify panel, right-click over an empty area of the panel (when you see the scroll hand), and from the pop-up menu select Display

17. In the Display rollout, check Hide Distribution Object

Scatter makes a clone of the distribution object (in this case the terrain). You don't need to see two terrains shown in the views, so you have just hidden the clone that Scatter made.

18. Use the right-click menu to scroll to the Scatter Objects area of the Modify panel

19. Set the Duplicates spinner to 50

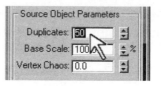

You should now see fifty trees scattered across the entire terrain. The duplicates are not standing straight up- they are at odd angles.

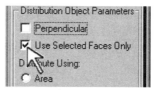

20. Uncheck Perpendicular, and check Use Selected Faces Only

If Perpendicular were left checked, each duplicate tree would align its Z axis with the Z axis of whichever face it was attached to, resulting in a forest of slanted trees. With Perpendicular unchecked, all the trees stand straight up.

With Use Selected Faces Only checked, the duplicate trees get scattered only over the faces of the terrain you selected with the Mesh Select modifier- the tree areas.

At this point all the tree duplicates are the same size and are rotated the same way. Use the Transforms rollout to add randomness to the size and rotation of the trees.

21. Use the right-click menu to scroll to the Transforms rollout

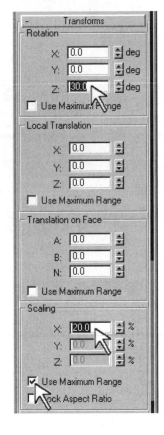

22. Type 30 in the Rotation Z: spinner, and Enter

The duplicate trees will be randomly rotated about their trunks, by as much as thirty positive degrees, and thirty negative degrees.

In the groups and spinners labeled Local Translation and Translation on Face, "Translation" just means movement. You can move each duplicate along any of its axes, and you can generate random movement of the faces of each duplicate along any of the axes of the faces.

23. In the Scaling group, set the X: spinner to 20, then check Use Maximum Range

The duplicates will be randomly scaled as much as 20% larger in all axes, and as much as 20% smaller.

Use Maximum Range is just a technical way of saying that 20% should be used for all three spinners.

To get just the right placement and composition of trees, try a few new random iterations of the scattering with the Seed button.

24. Near the bottom of the Modify panel, select the New button a few times, until the arrangement of the trees through the camera view is to your liking

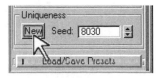

25. Save the scene, set the render type drop-down list (between the teapots) to View, and render the Camera view

These trees are quite convincing. They are also fairly "inexpensive," in terms of file size and rendering times. To get a mixed deciduous / conifer forest, your object to be scattered could consist of two crossed cutouts, one a fir tree and one a maple. The Transforms tools would insure that the distribution of firs and maples looked totally random.

Making Cutouts

The best way for you to make your own cutouts is in AutoCAD. Find a scanned photo of a tree against a black background (there are many such images available on the Web and on CD from various vendors). In AutoCAD, use the Insert / Raster Image tool to show the photo in the viewport. Trace the general shape of the tree with a Pline. Scale the Pline shape to the right height for the tree. Stand the shape up so the trunk is the Z axis (so that in a plan view it looks like a single straight line). Use a Move command to set the middle of the bottom of the trunk to location 0,0,0. Choose File / Export, and export as a .3DS file. Start VIZ, and import that .3DS file. Select the shape, apply a UVW Map modifier. Build a 2-sided material that uses the tree photo in the Diffuse Color channel, and a white-on-black version of the photo in the Opacity channel. Assign the material to the tree cutout plane. Copy the plane, rotate the copy 90 degrees, then use an Edit Mesh modifier to join both planes into a single cross-plane cutout. Collapse the Stack so the cutout is a simple Editable Mesh, and save the VIZ file.

Merge the House

The last item missing from this scene is the house, which has already been built and positioned, and can be merged in.

1. Close the rendered view

2. From the menus choose File / Merge

3. In the Merge File dialog box, browse to C:\Viztutorials\Chapter5, highlight HillHouse.max; choose Open

4. In the Merge dialog box, highlight both items in the list; choose OK

5. Save the scene and render the Camera view

Modeling Terrain Using Displacement

Remember the image you rendered and saved back on page 236? It's time to see what it can do before we leave the subject of landscapes. You're going to use that image as a Displacement Map. The image will be applied to a Patch Grid, which is like a sheet of rubber, and the lighter parts of the image will pull the land upwards while the darker parts push the land down. Displacement Maps work very well for gently rolling landscapes such as golf courses.

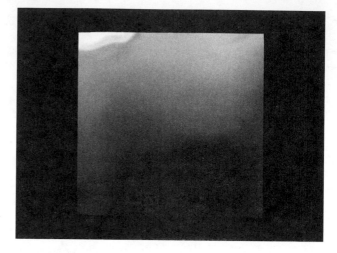

1. From the menus, choose File / Reset, and when asked "Do you really want to reset?," choose Yes

2. Make the Top view active

3. In the Create panel, Geometry category, open the subcategory drop-down list and choose Patch Grids

4. Choose Quad Patch

5. Open the Keyboard Entry rollout

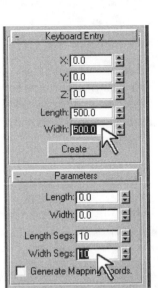

6. In Keyboard Entry, enter a Length and a Width of 500 feet

7. In the Parameters rollout, enter Length Segs of 10 and Width Segs of 10

8. Choose the Create button

9. Zoom Extents All

10. In the Perspective view, Arc-Rotate to get a good view of the Patch Grid

11. In the Modify panel, apply a Displace modifier

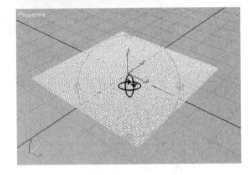

12. In the Image group, choose the button labeled None under the label Map: – **not** under the label Bitmap:

By assigning a map under the label Map, and not Bitmap, you are doing two things: you can choose any type of map (not just a bitmap ,but maybe a procedural map) and you are also setting up a live link to the material editor, so you can alter the way the map is used.

13. In the Material / Map Browser, highlight Bitmap, choose OK

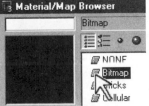

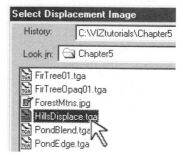

14. In the Select Bitmap Image File browser, browse to the C:\Viztutorials\Chapter5 folder, highlight the HillsDisplace.tga that you rendered earlier, and choose Open

15. Open the Material Editor. Drag the button labeled HillsDisplace.tga from the Modify panel and drop it onto an unused sample window. In the Instance (Copy) Map dialog box, select Instance, choose OK

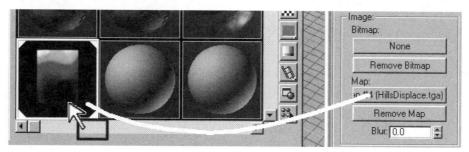

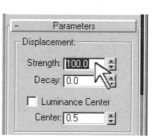

16. In the Modify panel, set the Strength spinner to 100

The displacement map has sculpted the landscape, but not correctly. The displacement map has a large black border area, which you can see is keeping the edges of the patch grid down at zero elevation. Crop the displacement map in the Material Editor.

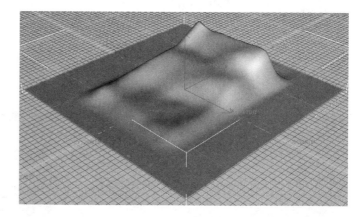

17. In the Cropping / Placement group in the Material Editor, check the box labeled Apply

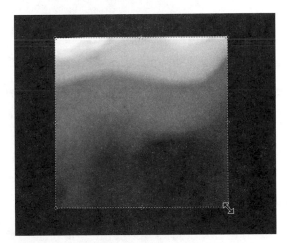

18. Choose the button labeled View Image

19. In the image view / crop window, drag the dashed cropping rectangle to crop to the displacement map. Crop off just a bit of the map on each side, to ensure that none of the border's black pixels will be allowed to pull the patch grid down at the edges. When the cropping rectangle is properly sized and positioned, close the view / crop window, and close the Material Editor

The displacement of the patch grid should be automatically updated in the views to show the effect of the cropping.

Using a displacement map on a patch grid produces a smoother landscape than the Terrain object. The landscape can be resculpted just by editing the bitmap in a paint program and reloading the bitmap in the Material Editor. As soon as you begin using Shapemerge to cut in landscape features, the object will no longer be a patch grid – it will become a mesh object, just like the Terrain object did when you Shapemerged into it. If you did not have any topographic information in CAD form for your site, but you had an aerial photo of the site, you could use that photo as the basis for creating a displacement map in a paint program, and then begin building your terrain model with a patch grid and the Displace modifier.

20. There is no need to save this file. Close VIZ without saving if you're done for the day, or Reset if you want to start the next chapter

Summary

You are probably relieved to see the "summary" heading, after seventy pages of landscape work. Hopefully the effort has convinced you that sitework and landscapes are not an insignificant part of a project, and if extensive landscape modeling is called for, budget plenty of time and brace yourself for what could be a major challenge. This exercise has obviously been set up to go well and to not overwhelm your computer's resources. In real projects you might not be so lucky. Remember to incremental save frequently; you should do this for all projects, but for scenes involving Terrain objects it is particularly important.

Keep in mind when modeling the terrain that fine detail is not necessary. A fairly coarse terrain can look great once good natural materials are applied to it.

Here is the finished image with the edges of the road roughed up in Photoshop using a Paint Daubs filter.

CHAPTER 6

Light and Shadow - Interior

If there is a Most Important Chapter in this book, this might be it. A suberb modeling job will result in a poor image if it lacks interesting, realistic light and shadow, and the simplest model can look great if properly lit. Lighting is challenging, particularly for interiors. There aren't many firm rules, and small changes can make large differences. Lighting is also time-consuming, as you cannot really know how you're doing until you render the scene, and well-lit interiors usually require sizable rendering times.

In this chapter you will open a scene of a tavern, with a bar, a pool table, dart boards, and some tables and chairs, and you'll light it. There is a good variety of lighting requirements in the tavern. There are pendant lights with globe-type diffusers that need light bulbs, there is the hood light over the pool table (which needs to show some smoke wafting in the light), there are wall sconces washing the corners and the ceiling with light, there are focused spots illuminating the dart boards, and there is a neon sign. The indirect, ambient light that bounces about in every real environment and creates secondary shadowing does not exist in VIZ, so you will need to simulate that effect.

General Approach

The best way to begin lighting an interior is to start with the feature lighting and end with general fill lighting. If you start the other way, by adding general illumination first, two things happen: first, you have a harder time seeing the effects of feature lights placed later because their effect is less dramatic in the already-lit environment, and second, you tend to over light the scene. Have you noticed a similarity among the interior scenes in this book? They are dark, moody scenes with high contrast. Look at magazines about 3D graphics, and look at Web sites and demo reels. They're filled with dark, evocative scenes populated with bright, shiny objects. That's not to say that brightly lit scenes are not exciting. They can be and should be, and you will have to learn to make them so. But in every project you do, you will be faced with the reality that the client or designer is more focused on an accurate depiction of the lighting in the space than with the creation of a knockout image. The point is that it is easier to make a darker scene really compelling, and as you work through this chapter, you may find that the point at which you are most satisfied with the look of the scene is just before you add the fill lights for general illumination.

Default Lighting

1. Open C:\Viztutorials\Chapter6\Tavern01.max

2. Quick-Render the Camera02-Panorama view

The scene is lit by two default lights. One is above the building, at the south-west, the other is below the building, at the north-east When you add the first of your own lights to the scene, these unseen default lights will turn off.

These two default lights do a decent job in providing general illumination, and although they will

not be used in this exercise, it is possible to add them to your lighting scheme. It is a two-step process. First, right-click over a view label (in the upper-left of each view), and from the menu choose Configure. In the Rendering Method tab of the Viewport Configuration dialog box, check Default Lighting, and 2 Lights.

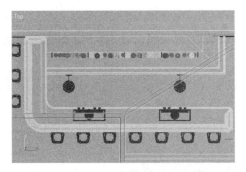

Then from the menus, choose View / Extra / Add Default Lights to Design

For this exercise, leave the default lights out of the scene. Let them turn off when you start adding lights.

Omni Light

Begin lighting the scene by adding lights to the two globe diffusers over the bar. These will be Omni lights, which are single-point light sources that cast light in all directions. They can be set to cast shadows or not. If shadow-casting is turned off, they illuminate everything, indefinitely.

1. Make the Top view active

2. Region Zoom to see the bar area

3. In the Create panel, Lights category, choose Omni

4. In the Top view, position the cursor over the center of the left globe diffuser, and click to set the light

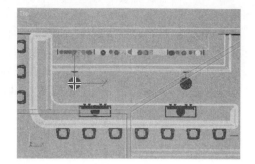

You probably won't be able to see the light in the Top view after you click; it is obscured by the globe, but you can see its axis icon.

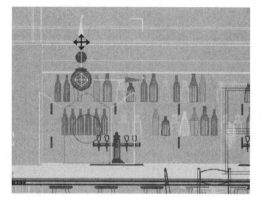

5. Make the Front view active, choose the Move tool, and move the light up off the floor, to the center of the globe (no need to be precise– close to the center will do)

6. Switch to the Modify panel

7. Name the light $Globe01

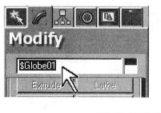

8. Set the light's Hue, Saturation, Value to:

    H = 40
    S = 15
    V = 255

9. Quick Render the Camera view

You can see two problems with this light. The first is most evident at the wall on the right- it's overly bright. The intensity of the light is not falling off with distance, that is, it is not attenuating. The second problem is with the shadows, and it is evident in the shadows of the bottles on the wall. The bottles are translucent and colored, but their shadows are opaque and gray.

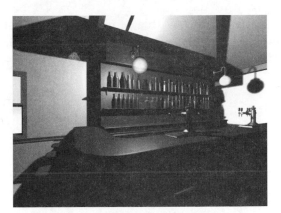

Deal with the attenuation first.

Attenuation

1. In the Modify panel, open the Attenuation Parameters rollout

There are two ways you can make a light's intensity fall off over distance: either with the Decay setting, or with the Attenuation values. You can change the Decay Type from None to either Inverse or Inverse Square. Set to Inverse, when the light has traveled four feet, it is one-fourth the intensity. Set to Inverse Square, when the light has traveled four feet, it is one-sixteenth the intensity. Inverse Square is closest to how light behaves in reality.

The Near Attenuation and Far Attenuation spinners offer complete control over intensity falloff. Near Attenuation is only occasionally used. It allows you to instruct the light not to begin illuminating until a certain distance away. If the Start spinner for Near Attenuation were set to 4 feet, and the End spinner to 8 feet, the light would cast no light for the first 4 feet. Between 4 and 8 feet, the light's intensity would grow, until at eight feet the light would shine at whatever intensity is set in the light's Value and Multiplier spinners.

With Far Attenuation, the light shines at full strength until it reaches the distance set in the Start spinner. It drops in intensity until it stops illuminating completely at the distance set in the End spinner.

Attenuation is activated by checking the check box labeled Use, for either Near or Far

305

Attenuation. With attenuation active, a selected light will show rings representing the start and end boundaries of the attenuation. With the Show box checked, a light will show the attenuation boundary rings even when the light is not selected.

 You can Non-Uniform Scale the attenuation rings. This may come in handy when you need to make an odd-shaped, special-purpose light.

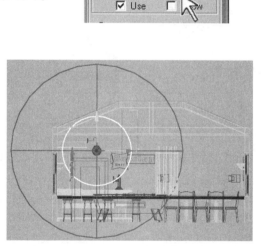

2. Check the Use check box for Far Attenuation

3. Set the Start spinner to 4′, and the End spinner to 10′

You now see two rings around the light. The Left view gives the best view of the falloff. The light will shine at full intensity on all the bottles. It will fall off just a bit as it gets to the bar surface, and it will barely reach the floor in front of the bar (and it will be very dim by the time it gets to the floor).

4. Save the scene into the C:\Viztutorials\Chapter6 folder. Name the scene Tavern*xx*.max, where *xx* are your initials

5. Render the Camera view

The attenuation is correct, now fix the shadows.

Shadow Maps Versus Ray Traced

The Omni light is by default set to use shadow maps. To reiterate the explanation of shadow maps in chapter 2 (page 81): use a mental image of the renderer looking down from the Top view, drawing a bitmap of the shadow shapes, and draping that bitmap down onto the scene at rendering. Shadow maps do not deal correctly with transparency. If you want shadows from the bottles that indicate the translucency and the colors of the bottles, you need ray traced shadows.

Ray traced shadows follow the path of light from the camera back to the source of light, calculating the effect on the qualities of the light as it bounces off or passes through various surfaces along the way. Why calculate from the camera to the light, when light travels the other way? Because the light from a light source travels out in many directions (in the case of an Omni, all directions), and it would take a very long time and be pointless to calculate all those rays when only some of them affect the scene, as seen from the camera's eye.

The advantages of ray traced shadows are the accuracy, and the ability to correctly interpret the effect of transparent objects. The disadvantage of ray traced shadows is that they take much longer to render than shadow maps, and that they cannot produce subtle, soft-edged shadows. It is not uncommon to use a combination of shadow maps and ray tracing to get the desired shadow effect. For example, in the case of the bar, you might set this first Omni light to shadow maps so it produces soft shadows from the bar back, the bar, and the beer taps, and you could tell this Omni to ignore the bottles. Then you would set another Omni light in almost the same place as the first and set it to cast ray traced shadows, and to only illuminate the bottles and bar back. The result would be soft shadows from the furnishings, and accurate shadows from the translucent, colored bottles.

Set the Omni light to cast ray traced shadows.

1. In the Modify panel, open the Shadow Parameters rollout

2. Open the drop-down list and change it to Ray Traced Shadows

3. Set the Density of the shadows to .7

With the shadows at Density of 1, the shadows from the bottles are so distinct that it is hard to tell them from the bottles, and the shadows everywhere else are too black.

4. Render the camera view

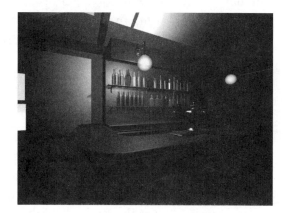

Much better shadows. On the ceiling you can see the shadow of the arm of the light fixture. This shadow would certainly be there in reality, but it would not be so distinct, as ambient light bouncing about the tavern would make it fuzzier and softer. This is a judgment call; what is going to be more noticeable- the hardness of that shadow, or the fact that it is not there if you remove it? Try removing it- you can always change back later if you want.

Object Properties-Cast Shadows

1. Type H, and select [GlobeLight01]. The name of the object is in brackets because the light fixture is several objects joined into a Group

2. From the menus, choose Modify / Properties

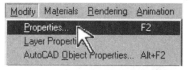

3. In the Object Properties dialog box, verify that the topmost button in the Rendering Control group is set to By Object. If it is set to By Layer, switch it to By Object

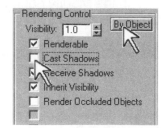

4. Uncheck Cast Shadows, and choose OK

5. Render the camera view. You don't need to let it render the whole scene. As soon as the ceiling is rendered, select Cancel in the Rendering dialog box

The ability to set any object to not cast shadows, not receive shadows, or both is a huge help.

Copy the Omni light to put a second one in the other globe diffuser.

Clone the Light

1. Make the Front view active

2. Type H, select $Globe01 (the Omni light)

3. Select the Move tool, position the cursor over the X axis of the Transform Gizmo, hold the Shift key on the keyboard, and drag to the right to clone a second light into the second globe diffuser

4. In the Clone Options dialog box, choose Instance, then choose OK. The cloned light will by default be named $Globe02

5. Zoom if needed to adjust the location of the light in the center of the diffuser

6. Type H, select [GlobeLight02] (the fixture)

7. From the menus, choose Modify / Properties

8. In the Object Properties dialog box, set the second light fixture not to cast shadows

9. Render the camera view

The scene is starting to look interesting. You can see that you're not going to need much fill light in this scene; in fact, you may decide you don't want any at all.

Later in this chapter you'll learn to use the Glow effect, which would help these globes look more convincing.

Chamfered Edges

This rendering is a good example of one of
the keys to good modeling, which is to add
chamfers to edges wherever appropriate.
Look at the valence of the bar back- there is
a gleam on the edge that would not be there
if that edge had been modeled as a hard
corner. Look at the monitor on the desk in
front of you. Is its form understandable to

you more because of its planar surfaces, or because of its edges? Some edges are picking
up the light, and some are creating shadow, and therefore contrast. These highlights and
contrasts are what your eye is drawn to as your eyes and brain interpret the object. Look
at the corners of the room you are in. Where the planes of the drywall meet, at both
inside and outside corners, do the seams appear to you to be darker than the planes?
They're not of course, but where two surfaces of contrasting value meet, your mind
perceives a dark stripe because of the contrast. You can model in such highlight and
contrast. As you model, envision the finished object and how the light will affect it, and
add chamfers or fillets wherever you think they will be effective. Make sure adjacent
wall surfaces are not too similarly lit. Get a good contrast between the planes so the
corners will seem better defined.

Spotlights

There are two track lights mounted to the ceiling above the dart boards. Set spotlights
into the track light fixtures.

1. Make the Camera02-Panorama view active

2. Right-click over the view label and from the menu
   choose Views / Camera01

3. Drag the Time Slider to frame 3

Before you render this view, you should turn off shadow-casting for the two Omni lights you just worked on, to speed up rendering as you work out the room's lighting scheme. Once all the other lights are in place and adjusted, you will turn on all shadow-casting for the final rendering.

4. Type H, select $Globe01

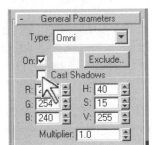

5. In the Modify panel, uncheck Cast Shadows

No need to select the other Omni light and turn off its shadow-casting; that has already happened, because you made the second Omni an instance of the first.

6. Render view Camera01 at frame 3

It is almost totally dark. There's some self-illumination in the materials for the stained-glass skylight, the wall sconces in the corners, and the bulbs of the track lights, and you can see the environment color through the window. In the real world, enough light from the two globes would bounce around the space to provide at least some illumination to the whole room, but VIZ light does not bounce.

7. Make the Left view active, and zoom in on the right-hand side of the building, where the dartboards and track lights are

8. Type W to maximize the view

9. In the Create panel, Lights category, select Target Spot

311

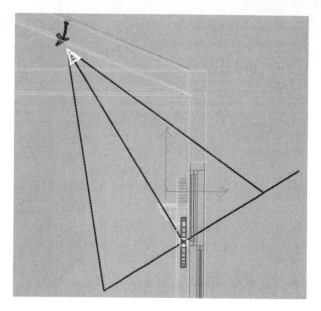

10. Position the cursor a bit in front of the track light, click-hold, drag the light so that the target is at the center of the dartboard, and release

11. In the Create panel, name the light $Darts01

12. Toggle to four views, make the Top view active, and zoom to see the light and the two dartboards

13. Type H, select both the light and its target

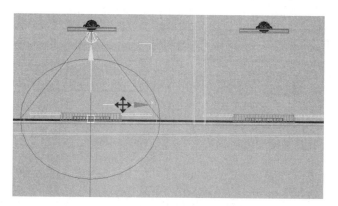

14. Select the Move tool and use the Transform Gizmo to move the light and target to be centered on the left-hand dartboard

15. Shift-Move the light and target to clone a second spotlight aimed at the right-hand dartboard. Choose clone type Instance

16. Render the camera view

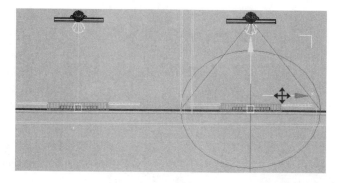

Hotspot and Falloff

The extreme focus of the spotlights is because their Hotspot and Falloff values are very close together. Hotspot and Falloff are a type of attenuation. Except in this case, the falloff is not measured in distance, but in angle, and it is measured perpendicular to the light vectors, not along them. Hotspot and Falloff is sideways attenuation, measured in degrees.

1. Type H, select $Darts01

2. In the Modify panel, set the light's HSV spinners to:
    H = 140
    S = 15
    V = 255   the light will be very pale blue.

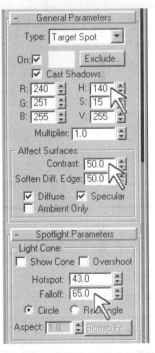

3. Set the Contrast spinner to 50
    This is a way of giving a light more "punch" without making it brighter. Overly bright lights cause colors to wash out.

4. Open the Spotlight Parameters rollout

5. Set the Falloff spinner to 65

The larger cone now fills 65 degrees

6. Render the camera view

With some separation between Hotspot and Falloff, you get softer pools of light. The light is at the same intensity (value), but seems brighter because the Contrast value of 50 is making any materials the light falls upon show more contrast.

7. Save the scene

Shadow Parameters

These two spotlights would, in reality, cast fairly crisp shadows, but probably not quite the knife-edge ones shown in the last rendering. The lights use ray traced shadows by default. Switch to shadow maps.

1. In the Modify panel, open the Shadow Parameters rollout

2. Open the drop-down list and choose Shadow Map

3. Set the shadow Density to .85

4. Open the Shadow Map Params rollout

5. Set the Sample Range to 8

See Chapter 3, page 152 for the explanation of Bias, Size, and Sample Range

6. Render the camera view

Increasing the Sample Range has softened the shadows just a bit.

You might try adding Far Attenuation to these spotlights – without it the light is a bit much on the floor. The image shown here uses Far Attenuation with a Start value of 9′, and an End value of 14′.

Wall Sconce

There will be three wall sconces in the corners of the room. You will set one up and clone it to two other corners. What you will find if you try to set up a wall sconce realistically, with the light source positioned in the sconce fixture, is that it doesn't do much. In reality, wallboard is not smooth. It has small bumps that can pick up light traveling nearly parallel to the plane of the wall. A VIZ wall doesn't have bumps, and if you shine a light nearly parallel to the wall surface, the angle of incidence of the light rays to the wall allows almost no illumination. Light objects for wall sconces need to be placed a foot or two away from the wall, aimed at a point on the wall above the sconce fixture.

1. Make the Top view active, maximize the view, and zoom to a close-up of the lower-left corner of the room, where the wall sconce is mounted

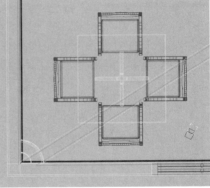

2. In the Create panel, Lights category, select Target Spot

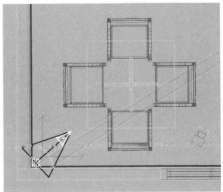

3. Drag a Target Spot light, starting a couple feet from the corner, and releasing the Target at the inside corner of the walls, as shown here:

4. Name the light $Sconce01

315

5. Toggle to four views

The placement of the light and its target needs to be exact for the light to throw the right shape pool of light up the wall and onto the ceiling. The pool of light needs to appear to begin just at the top of the sconce fixture, and it needs to open wide as it travels up, to illuminate a large area of the corner and ceiling. There is no science to getting the light positioned correctly, it requires trial and error. To spare you the tedium of moving the light an inch, rendering, moving the target an inch, rendering, and so on, use the Transform Type-In to position the light and target precisely.

6. With the light still selected, select the Move tool and right-click over it to open the Transform Type-In

7. In the Absolute: World spinners on the left, enter values of:

   X = -22'11" (note the minus)
   Y = -11'5" (note the minus)
   Z = 6'8.5"

Leave the Transform Type-In open

8. Hit Esc on the keyboard to make sure you are no longer typing in one of the spinners, then Type H, select $Sconce01.Target

9. In the Transform Type-In, in the Absolute: World spinners, enter values of:

   X = -23'11" (note the minus)
   Y = -12'4" (note the minus)
   Z = 10'

10. Close the Transform Type-In

11. Make the camera view active, make sure this view is still showing Camera01, and make sure the animation is still at frame 3

Rather than render the whole view, which shows more of the room than you're concerned with at this point, just render an enlarged portion of the view, with the Render Blowup option.

Render Blowup

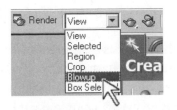

1. Open the render type drop-down list, between the three teapots, and choose Blowup

2. Choose the middle teapot – quick render

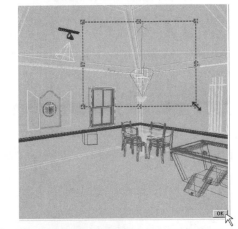

3. A dashed rectangle appears in the camera view. Resize it by its corners and move it by dragging when the cursor is inside the box, showing a four-way arrow. Position it as in this figure, then select the OK button in the lower-right of the camera view (remember that the OK button tends to disappear – if you do not see it, just click where you know it should be)

Doesn't look like much yet, but it will once you set the Hotspot and Falloff values for the light.

Spotlight Parameters

1. Type H, and select $Sconce01

2. In the Modify panel, set the light's HSV spinners to:

> H = 255
> S = 15
> V = 255

This light will have a slight pinkish cast.

317

3. Set the light's Multiplier to 1.25

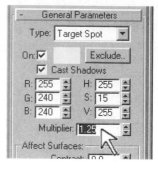

The Multiplier is a dimmer switch for lights. Generally, it is used in two situations: if a light's Value is at 255, which is as high as Value goes, and the light is still not bright enough in the rendering, you can make it brighter with the Multiplier. Its other use is in animations. It is an animatable value. You can turn on the Animate button, set the Multiplier to 0 at frame 1, move to frame 50, set the Multiplier to 1.0, and the result will be that the light ramps up from darkness to full intensity over fifty frames.

In the case of this sconce, you need a bit more intensity than the maximum 255 Value, because the angle of incidence of the light rays to the wall surface is so low. The Multiplier value of 1.25 will help the light illuminate the walls sufficiently.

⚠ You can set the Multiplier to a negative value to create a light that darkens, rather than illuminating.

4. In the Affect Surfaces field, set the Contrast spinner to 0

5. Open the Spotlight Parameters rollout

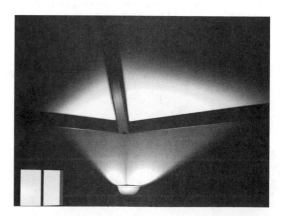

6. Set the Hotspot spinner to 125

7. Set the Falloff spinner to 150

8. Choose Render Last

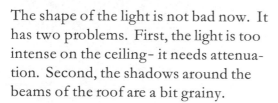

The shape of the light is not bad now. It has two problems. First, the light is too intense on the ceiling- it needs attenuation. Second, the shadows around the beams of the roof are a bit grainy.

Attenuation

Because the angle of incidence of the light to the ceiling is much more perpendicular than the angle of incidence of the light to the wall, the ceiling is more strongly illuminated, and is overlit.

1. Open the Attenuation Parameters rollout

2. Set the Decay Type drop-down list to Inverse

3. Set the Decay to start at 4'

4. Render Last

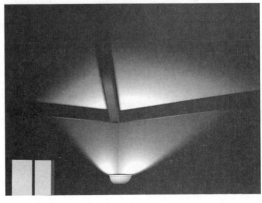

Why was Inverse Decay chosen, as opposed to using Far Attenuation and adjusting the Start and End values? How do you know which type of attenuation to use in which situation? You don't. It is just trial and error, and you will have to try the choices in every situation.

Shadow Map Settings

1. Open the Shadow Parameters rollout and verify that this light is set to use Shadow Maps

2. Set the shadow Density to 1.0

3. Open the Shadow Map Params rollout

4. Set the Map Size to 800

5. Set the Sample Range to 14

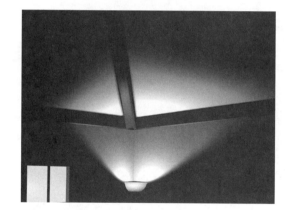

6. Render Last

There is now better definition at the edges of the shadows, because of the higher map size, and the shadows are a bit broader, due to the higher Sample Range.

Group the Light and Fixture

You need to duplicate the sconce into two other corners of the room. This will be much easier if the fixture and the light object are grouped together, to maintain their precise placement relative to each other.

1. Type H, select $Sconce01, $Sconce01.Target, and [Sconce01]

2. From the menus, choose Draw / Group / Create

3. In the Group dialog box, name the group SconceGroup01

320

Clone the Sconce

1. Make the Top view active, maximize it, and Zoom Extents

2. With SconceGroup01 selected, choose the Mirror tool from the vertical toolbar at the left of the views

3. In the Mirror dialog box, set the Mirror Axis to X

This is the axis **along** which you want to mirror, not the axis **about** which to mirror, as in AutoCAD.

4. Set the Clone Selection to Copy

You may find a unique lighting challenge in one of the corners of the room, so you want each clone to be independently editable.

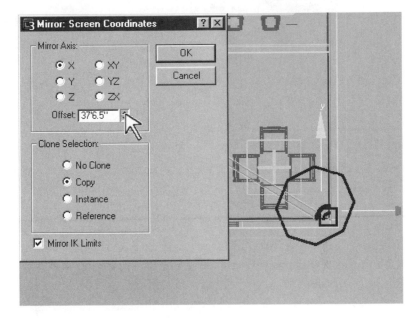

5. Drag upwards on the small up arrow at the right of the Offset spinner. The new light will start to move to the right, slowly. To speed up the movement, hold the Ctrl key on the keyboard as you drag the spinner. Position the new light by eye in the lower-right corner of the room, as shown in the preceding figure, and release the mouse

6. Choose OK to close the Mirror dialog box

7. Type H, select SconceGroup01 again

8. Use the Mirror tool to clone the light into the upper-left corner of the room. This time the Mirror Axis will be the Y axis

9. Once two new lights have been made, zoom tight to each new light and fine-tune its position in the corner

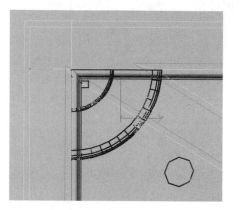 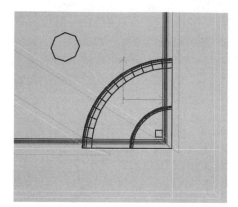

10. Save the scene

Sunlight System

VIZ's sun-angle calculator can tell you exactly what time, on any given day, your high-rise building design will send the neighbors into shade, and it can tell you how to configure overhangs and windows and place furnishings to provide your interior with warming sun in the winter and cooling shade in summer. While the daylighting performance of a building is usually not the first priority among the patrons of a tavern, you'll add late-morning sun through the southern windows of this pub by way of introduction to the Sunlight System.

1. Zoom Extents, then choose the Zoom tool

2. Place the cursor at the top of the view, and drag downward. Starting the drag at the top of the view will reveal the area south of the building. Zoom until the building takes up only about a third of the view

3. In the Create panel, Systems category, select Sunlight

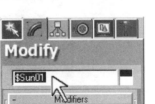

4. Place the cursor at the center of the building, and drag out a compass rose, about the size of the building. Release the mouse and move the mouse downward in the view. You'll see a light object moving away from the building. Place the cursor at the very bottom of the view and click to set the position of the light

Your light will not be at the same position as what is shown here. It will be set to the position of the sun at whatever time and place you are as you do this exercise.

5. Switch to the Modify panel

6. Name the light $Sun01

7. Set the light's HSV values to:

    H = 158
    S = 16
    V = 180

Assume it is an overcast day, so the sun has a bluish cast to it.

8. Set the Multiplier to 1.0

9. Open the Directional Parameters rollout

10. Uncheck Overshoot

11. Set Hotspot to 25', Falloff to 27'

Overshoot causes light to be cast beyond the
boundary of the light shown by the blue
cylinders in the views. This prevents a giant spotlight from appearing on the ground.
Although light will be cast indefinitely with Overshoot on, shadow will not; shadows will
only appear within the boundary defined by the blue cylinders.

12. Switch to the Motion panel

The parameters for a Sunlight System appear in two panels- the Modify and the Motion.
The Modify panel carries the parameters of the light itself, while the Motion panel deals
with the path of the Sun in the sky for whatever location and time you choose.

13. Select the Get Location button

14. Scroll the list and select
Boston, MA; choose OK

15. Set the time and date:

Hours = 12, Mins = 0, Secs = 0

Month = 10, Day = 1, Year = 2000

Verify that Daylight Savings Time is
checked

16. Set the Orbital Scale to 75'

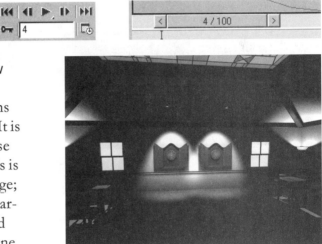

Orbital Scale is the distance from the compass rose to the light. A rule of thumb for the correct distance is roughly three times the length of the object the light is illuminating.

Note the North Direction spinner. If your plans are not drawn oriented north, use this spinner to rotate the compass rose so it points north.

17. Restore four views, make the Camera01 view active, and move to frame 4 of the animation, by either typing 4 in the frame field of the time controls, or by moving the Time Slider below the views

18. Set the render type drop-down list (between the teapots) to View, and render the camera view

The shadows from the window mullions and the furniture are too hard-edged. It is supposed to be a bit hazy out, and these shadows indicate a clear blue sky. This is another example of a common challenge; only ray traced shadows respect transparency (the window glass), and ray traced shadows are often too crisp for the scene. The only way to get a bit of blur on these shadows from the windows is to switch to shadow maps. But the shadow maps will not go through the transparent glass in the windows. The solution isn't really that difficult.

VIZ Window objects are modeled with the glass being made of two planes. One plane's normals point outside the building, the other plane's normals point inside the building (see Chapter 2, page 54, for an explanation of normals and backfaces). If you delete the outside-facing plane, then the light shining from outside only meets the plane that is facing away from the light source – the light does not see that plane, and the light and shadows are allowed to travel into the room.

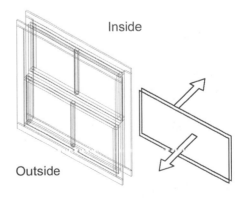

Inside

Outside

Edit Mesh – Delete Faces

1. Make the Left view active and maximize it

2. Arc-Rotate, Pan and Zoom to get a good view of the south side of the building (the wall with the dartboards)

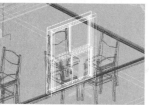

3. Type H, select SlidingWindow03

4. Zoom Extents Selected

5. Save the scene

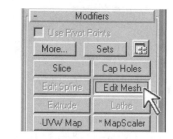

6. In the Modify panel, apply an Edit Mesh modifier to the window

7. Switch to Sub-Object Polygon

8. Position the cursor on the upper sash, where glass would be, and click. One polygon will highlight red. Hold the Crtl key on the keyboard and click to select the polygon of the lower sash's glass.

   With both outside polygons of the glass selected, hit Delete on the keyboard

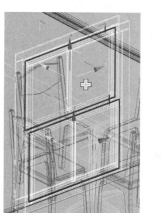

9. Turn off Sub-Object

10. Type H, select SlidingWindow04, and repeat steps 4 through 8 on the previous page to remove the outward-facing polygons of the glass for that window

11. Type H, select $Sun01

12. In the Modify panel, Shadow Parameters rollout, switch from Ray Traced shadows to Shadow Map

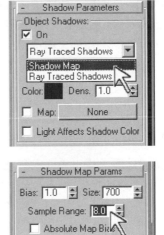

13. Set the Map Size to 700 and the Sample Range to 8

14. Toggle back to four views

Render Region

To see the effect of the light through the windows casting shadows on the furnishings and floor, why render the entire view, when you're only working on the lower third? Use Render Region to render just the lower third.

15. Make the camera view active

16. Between the three Render teapots, open the drop-down list and select Region

17. Select Quick Render

18. A dashed rectangle appears in the camera view. Resize and move it so it surrounds the floor and tables, as shown here:

19. Select the OK button in the lower-right of the camera view

Hmmm. . . no light, no shadows.

There's one thing left to do to make this strategy work. The premade materials for Door and Window objects that are in the libraries all specify that the glass should be 2-Sided. A 2-Sided material shows up on both sides of a face– both the normal and the backface. This means that the Sun light is "seeing" the remaining glass polygons in the windows, even though those polygons face away from the light. By turning off the 2-Sided property in the glass material for these windows, the light will then not see the glass polygons, and you will have sunlight and shadow in the pub.

Material: Turn Off 2-Sided

1. Open the Material Editor

2. Make the upper-left sample window active

3. This first sample window should show the material Window–Oak / Clear. If it does not, choose the eye-dropper button just to the left of the material name and in the camera view, pick on any part of a window to load that material into the editor

4. This material is a Multi / Sub-Object material. Select the Glazing sub-material to drop to that level of the Material Editor

5. In the Shader Basic Parameters rollout, uncheck 2-Sided

6. Close the Material Editor

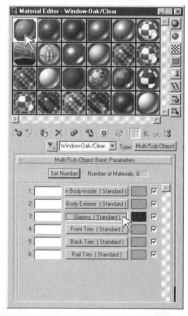

7. Select the rightmost teapot, Render Last

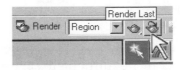

Render Last will render the same view you rendered previously, using the same settings as before, and if you used Render Region, or Render Blowup, or another choice from the drop-down list, it will honor that choice as well. Quick Render, on the other hand, renders the **current** view, using the same settings as the previous render. If you don't want to define the Region window again, or if the camera view is no longer the active one, use Render Last.

Now you're getting somewhere. The shadows on the furnishings and on the floor are less distinct, and look more believable for a hazy day. There's just one more problem. Look closely at where the wall meets the floor. There is a strip of light there, as if light is leaking in at the bottom of the wall. This is because the Bias setting for the shadows is too high.

Map Bias

Remember from Chapter 3, page 152, that Bias is like a rubber-band connecting the shadow to the object. If the Bias is too high, the rubber-band is too long, and the shadow separates from the object (that's the case in this scene). If the Bias is too low, the shadow can actually begin in front of the object. Set the bias down to pull the wall's shadow back so it meets the wall correctly.

1. Select $Sun01, if it is not already selected

2. In the Modify panel, Shadow Map Params rollout, set the Bias to .25

3. Render Last

The light at the bottom of the wall should be gone.

4. Save the scene

Volume Light

A Volume light is a light with particles floating in it such as dust, smoke, and fog. Volume lights are not difficult to set up, and they do a great job in adding mood to a scene, but they also can increase rendering times dramatically. Add a light source to the light hood over the pool table, and add some smoke to the pool of light.

1. Make the Top view active, and zoom in on the pool table

2. In the Create panel, Lights category, select Free Spot

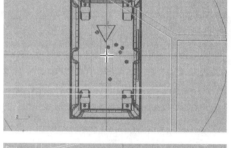

3. Place the cursor over the center of the pool table in the Top view, and click to create the light

If the light appears in the views not as a cone as shown in these images, but as an extremely wide cylinder, don't be alarmed. You will correct this in a moment.

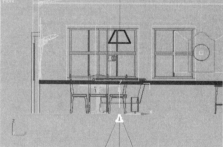

4. Make the Front view active and zoom in on the light, the pool table, and the hood light

5. Move the spotlight up. Position the light a bit above the hood light, as shown here:

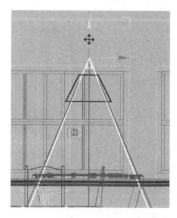

6. In the Modify panel, edit the light's parameters:

    Name the light $Pool Light

    Make sure Cast Shadows is checked

    Set the HSV spinners to H = 40, S = 25, V = 255

    Set the Multiplier to 1.5

    In the Affect Surfaces field, set Contrast to 25

In the Spotlight Parameters rollout:

    Set the Hot Spot to 55, and set the Falloff to 66

    Set the light's shape to Rectangle

    Set the Aspect of the rectangle to .5

    Set the Target Distance to 9'4"

In the Attenuation Parameters rollout:

    Set the Far Attenuation spinners to: Start = 8' End = 14'

    Check the Use check box for Far Attenuation

In the Shadow Parameters rollout:

    Set the shadow type drop-down list to Shadow Map

    Set the shadow Density to .9

In the Shadow Map Params rollout:

    Set the shadow map Bias to 1.0, the Size to 512, and the Sample Range to 12

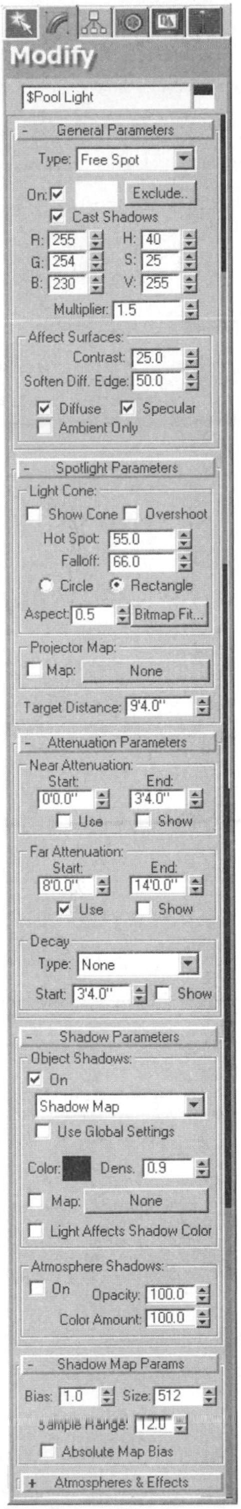

7. Working in the Front and Left views, move the light so that the Falloff cone (the outer, darker one) fits the bottom of the light hood, as shown below:

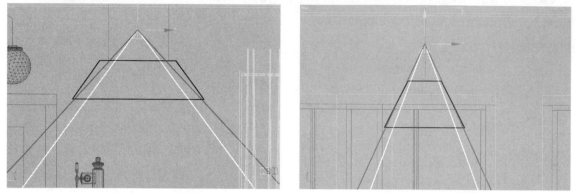

If you were to render the pool table area at this point, you would see no light falling on the pool table and the floor below. This light will reach the top of the light hood and stop. So it will actually be putting the pool table in shadow, not light. You need to exclude the light fixture from being seen by the light.

Exclude / Include

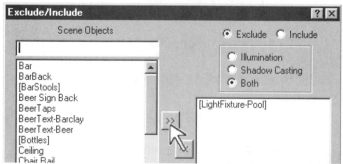

8. In the Modify panel, select the Exclude button

9. In the Exclude / Include dialog box, highlight [Light Fixture-Pool] (in brackets because it is a group), then select the right-pointing arrow to transfer the fixture to the exclusion list. Choose OK

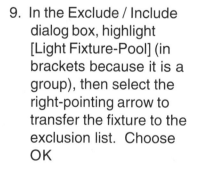

10. Make the camera view active. This should still be the Camera01 view. Set the Time Slider below the views to frame 0, and render the camera view

The light hood glows inside because of Self-Illumination. The outside of the hood is dark metal, while the inside is a yellow-ish self-illuminated material. You will learn to make such a

332

material in the next chapter.

This light's illumination is OK, it just needs some smoke floating in it.

11. Save the scene

12. At the bottom of the Modify panel, open the Atmospheres and Effects rollout

13. Choose Add

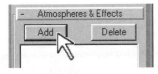

14. In the Add Atmosphere or Effect dialog box, select Volume Light; choose OK

15. In the list of Atmospheres and Effects at the bottom of the Modify panel, highlight Volume Light, and choose Setup

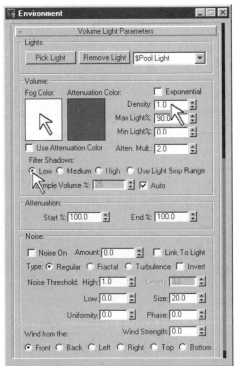

16. In the Volume Light Parameters rollout, select the Fog Color swatch, and set the fog color to H = 40, S = 15, V = 255

17. Set the Density to 1

18. Set Filter Shadows: to Low

Of all these parameters, the Density value has the greatest effect on the appearance of the particles in the light. The correct density is different for each situation; it is affected by the broadness or narrowness of the pool of light, the distance from the light to the camera, and a few other variables. Find the right setting through trial and error.

Filter Shadows determines the quality of shadows that objects in the fog will cast on the fog. The Use Light Sample Range setting means that if the light has a large Sample Range in its shadow map controls, the shadows in the fog will be soft-edged, and if the light has a small Sample Range, the shadows will be crisp. If your scene features objects in the fog that need to cast shadows that match the rest of the scene, this is probably your best setting. For this scene, the only object casting shadows in the fog is the pool table, and it is hardly noticeable, so it's best to use the Low quality setting to save rendering time.

19. Leave the Environment dialog box open, and use Render Region to render just the area above the pool table

It's a nice effect, but it might be nicer if the haze was not so uniform. Try adding a bit of noise to the effect.

20. In the Noise group of the Volume Light Parameters, check Noise On

21. Set the Noise Amount spinner to .2

22. Set the Noise Type to Turbulence

The Noise will make the Volume Light effect a bit more pronounced, so you need a bit less density to keep the effect subtle.

23. Set the Density to .8

24. Render Last

The light looks smokier now.

There is one more thing to fix. Look
closely above the light hood- the cone of
the volume light begins above the light
fixture. You can barely see it in this scene,
but in cases where the correct position for
the light is further above the fixture, it will
be very obvious. The light needs Near
Attenuation, that is, it needs to cast no
light until the cone of light is within the
hood.

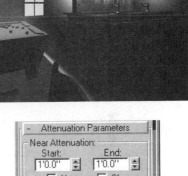

25. In the Modify panel, Attenuation Parameters
    rollout, check the Use check box for Near Attenua-
    tion, and set both the Start and End spinners to 1'

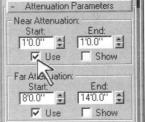

26. Use Render Region to render a small area above
    the hood, and see if you can detect the volume
    light disappearing as the region is re-rendered

You must have noticed that rendering was slow in the area of the pool table, once you
created a volume light. It is a somewhat expensive effect, and even more so if the scene
also involves objects within the fog, casting shadows, and reflective surfaces that need to
show the volume light in the reflection.

Because it takes so long to render, you probably will want to turn the volume light effect
off, once the parameters are correct, so that as
you work on the rest of the lighting and
materials, your rendering times won't get out
of hand. There are two ways to do this. In the
Atmoshpere rollout of the Environment dialog
box is a check box labeled Active, which you
can uncheck until final rendering time.

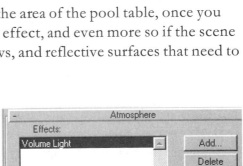

27. Save the scene and close the Environment dialog box

The Render Design dialog box has a check box labeled Atmospherics, that when unchecked will prevent the rendering of any volume lights, ground fog, background fog, and combustion.

Options:
- ☐ Video Color Check ☐ Atmospherics ☐ Super Black ☐ Render Off
- ☐ Force 2-Sided ☑ Effects ☑ Displacement ☐ Render to Fields

Multiple Volume Lights

The easy way to make several volume lights (a row of street lamps, for example) is to set one up and clone it. If you need control of each volume light's parameters individually, you will need to set each light up individually, using steps 12 through 15 on page 333 to assign a volume light atmosphere and access the parameters.

Adding multiple volume lights to a scene can also be done entirely in the Environment dialog box. Suppose you wanted three distinct looks to volume light effects in this tavern scene: the subtle smoke in the pool table light, a halo around each of the globe lights over the bar, and some dust in the beams of light coming through the windows. This would be the procedure (just read these steps, don't do them):

In the Environment dialog box, Atmospheres rollout, choose Add/Volume Light two times to add two more effects to the list.

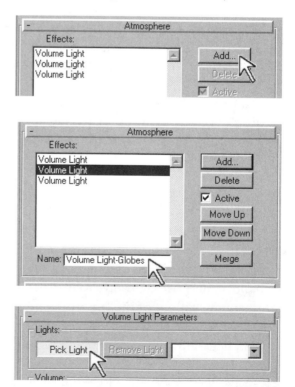

Highlight the second Volume Light effect in the list to access its parameters, and name it Volume Light-Globes (you should also rename the first one Volume Light-Pool Table).

Choose the Pick Light button, type the letter H, and from the list of lights, choose $Globe01. Then repeat this procedure and pick $Globe02.

Now, in the drop-down to the right of the
Pick Light button, both $Globe01 and
$Globe02 would be listed.

Set up all the parameters and do all the
test renders until the halos around the globe lights show the desired effect.

Highlight the third Volume Light effect in the list, name it Volume Light-Sun, choose
Pick Light, type H, and pick $Sun01 from the list. Set up all the parameters to make the
dust dancing in the light streaming through the windows look just the way you want.

Neon: Glow Effect

There are two parts to a neon effect. First you need to make the tubes appear brightly
illuminated, and second, you need to simulate the light from the tubes falling on objects
nearby. The illumination of the tubes is done in the tubes' material, using self-illumination. An object carrying a self-illuminated material does not cast light. The simulation
of illumination from the tubes falling on nearby objects is most easily accomplished with
a Glow effect, which is added to the image by the renderer in a second rendering pass.

Glow is one of several effects in a category called Lens Effects, so named because they
work when you render through a Camera view. Other effects in this category add lens
flare to an image, or star-shaped highlights glinting off extremely shiny objects, or rings,
like the ring around the moon on a hazy night.

There are three ways to add a Glow effect to a scene. You can assign a light to cast a
glow effect, you can assign a material to use Glow, or you can assign an object to cause
Glow. In the case of neon, either the second or third techniques will work. The easiest
way is probably to use by-object assignment.

1. Use the Time Slider or the Time Controls to move to frame 5 of the animation

Since you're going to be working just with the neon sign for a few minutes, turn the display of everything else off to speed up viewport and rendering speeds.

2. From the menus, choose Tools / Floaters / Display Floater, and in the Display Floater, choose the Off / By Name button

3. In the Turn Off by Name dialog box, from the buttons below the list of scene objects, choose All

4. Hold the Ctrl key on the keyboard and from the list deselect Sign Back, Sign Text-Barclay, and Sign Text-Beer

5. Choose the Off button

6. Move the Display Floater out of the way

7. Render the Camera01 view

The self-illuminated material has already been developed and applied to the neon tubing. You'll learn to make self-illuminated materials in the next chapter.

In the views, the neon looks like splines, but in the rendering it looks like tubes with thickness. Select just one piece of text and look in the Modify panel to see how this is done.

8. Type H, select Sign Text-Barclay

9. In the Modify panel, open the General rollout

Renderable Splines

The Renderable check box is checked, and the render Thickness is set to .5″. Making splines renderable is the way to avoid having to model things such as ornate railings, parking lines on pavement, a coil of rope, and power cords from a computer. You just draw the object as a spline, tell it to be renderable, and the renderer will interpret the splines as tubes. These neon lights were created with the Text tool in the Create panel, Shapes category.

Renderable splines do not have to render as tubes. To flatten a renderable spline, select the spline, make the Non-Uniform Scale tool active (it is nested in the flyout with the Scale tool), then right-click over the Non-Uniform Scale tool. In the Offset:World Z spinner, enter 0. The spline will render as a ribbon, not as a tube. This trick only works on 2D splines. If you try it on a 3D spline, you will change the Z values of the various vertices of the spline, as well as changing the way it renders. But it works perfectly for the lines on a flat parking lot.

G-Buffer / Object Channel

To make an object show the Glow effect, you assign a number to the object's effects channel, called the G-Buffer channel, then you set up the effect parameters and instruct the effect to apply itself to any objects carrying that G-Buffer number.

1. With Sign Text-Barclay still selected, choose, from the menus, Modify / Properties

2. In the Object Properties dialog box, set the G-Buffer / Object Channel spinner to 1

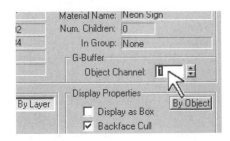

3. Close the Object Properties dialog box

4. Type H, select Sign Text-Beer, and repeat steps 1 through 3 to also assign G-Buffer Channel 1 to that neon tubing

5. From the menus, choose Rendering / Effects

339

6. In the Rendering Effects dialog box, choose the Add button

7. In the Add Effect dialog box, highlight Lens Effects; choose OK

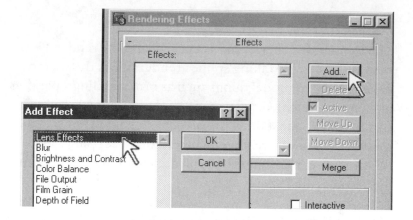

8. In the Lens Effects Parameters rollout, highlight Glow in the list at left, then choose the right-pointing arrow to add Glow to the list at right

9. Highlight Glow in the list at right to access all parameters for that effect

There are two rollouts containing settings for Lens Effects: the Lens Effects Globals rollout, and the Element rollout. Settings in the Globals rollout pertain to any Lens Effects you add to your scene, while settings in the Element rollout pertain only to one particular effect.

If you were to add three Lens Effects to the scene, and all three were to appear fairly small and not too intense in the image, you would want to set the Size in the Globals rollout small, maybe 10, and the intensity in the Globals lower. Then you would adjust the size and intensity of the individual effects in the Element rollout for each effect.

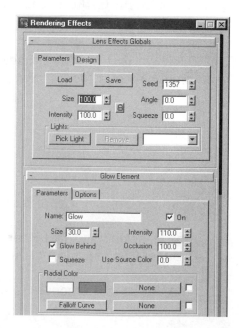

10. In the Glow Element rollout, choose the Options tab

11. Put a check next to the Object ID spinner, and verify that the spinner is set to 1

If the Glow was determined by material, you'd check the Effects ID spinner, and set that spinner to whatever number you had assigned to the Effects channel within the material.

12. In the Glow Element rollout, select the Parameters tab

13. Name the effect Glow-Neon Sign

14. In the Lens Effects Globals rollout (not the Glow Element rollout), set the Size spinner to 5

15. Render Last

After the initial rendering pass, a second pass will add the Glow effect

You don't need to re-render the scene every time you want to see the effect of changes to settings in the Glow parameters. Because the Glow is added in a second rendering pass, you can just update that second pass to quickly see the effect of parameter changes.

16. Size the Virtual Frame Buffer to a window just large enough to see the sign rendered, and use the scroll bars in the VFB to pan the sign into the resized window. Then move the VFB so you can see it and the Rendering Effects dialog box side-by-side

The Glow effect is too broad, too intense, and too white.

17. In the Glow Element rollout, set the Size spinner to 2, and the Intensity spinner to 90

18. Scroll up the Environment dialog box and choose Update Effect

The Glow effect will update in the VFB without a full scene re-render.

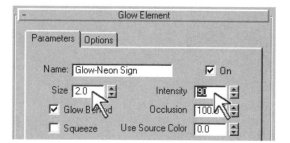

342

The Glow is now a good size and intensity, and just needs to be more yellow.

19. Scroll back down the Environment dialog box to the rollouts. You won't be changing any more settings in the Lens Effects Parameters rollout or the Lens Effects Globals rollout, so roll them shut to prevent unnecessary scrolling

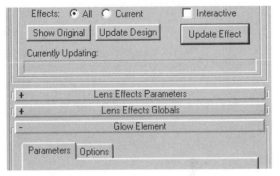

20. In the Glow Element rollout, select the white color swatch in the Radial Color group, and set the color to:
H = 40, S = 100, V = 200

21. Choose the Update Effect button again

You should now have a good yellow neon glow around the letters.

22. Save the scene

23. Close the Rendering Effects dialog box and the VFB

Lens Effects settings can be saved and recalled later. In the Lens Effects Globals rollout are the Load and Save buttons. The Load button directs you to the Plugcfg folder, where a few settings files are available for loading. You'll find one for a candle glow in the Chapter 1 folder for this book.

Fill Lights

You are done with the feature lights in this scene. Just a few more lights need to be added– a couple to provide general illumination, and a couple that will add no illumination, but just create additional shadowing throughout the tavern. Add the two fill lights.

1. In the Display Floater, choose On / All

2. Set the four viewports to Top, Front, Left, and Camera01, if they are not already, and Zoom Extents All

3. Set the animation to frame 0

4. Render the camera view

Notice the neon sign Glow effect– it appears to have grown. This is because the Size values in the Lens Effects settings are a percentage of the size of the rendered image. Up close to the neon sign, the 2% value you set in the Glow Element rollout was OK, but in a more distant view, it is too large. Unfortunately, this is not an animatable parameter in VIZ. If you have a Glow effect occurring over several different views, you will probably have to test the Size parameters of the Glow and adjust them before rendering each view.

5. Close the VFB and make the Front view active

6. Zoom out to see more space above and left of the building

7. In the Create panel, Lights category, select Omni

8. Create the light left of the building and about as high as is shown in this image:

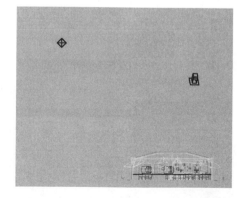

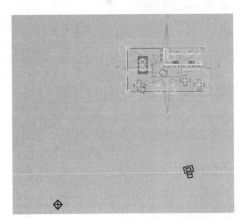

9. Switch to the Top view, and move the light so it sits off the southwest corner of the building, as shown here:

10. In the Modify panel, set the parameters for the light:

Name the light $Fill-SW

Uncheck Cast Shadows

Set the HSV spinners to:
H = 0, S = 0, V = 140

Set the Multiplier to 1.0

Set Affect Surfaces / Contrast to 0

11. Pan and zoom the Top view so you have some room off the northeast corner of the building

12. Shift-move $Fill-SW to clone another fill light northeast of the tavern. Choose clone type Copy, and name the new light $Fill-NE

Consider the angle of incidence of the rays from each

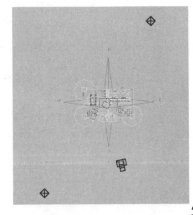

345

light onto the wall surfaces. You want contrast between adjacent walls, so the lights should not be aimed exactly along the diagonal. Aim them so each light will illuminate a long wall section a bit more strongly than a short wall segment.

13. Move the lights in the Top view to make them illuminate the longer walls a bit more than the short ones

14. Render the camera view

The only part of the tavern not sufficiently illuminated is the ceiling. Put one more light below the building for the ceiling.

15. In the Top view, create another Omni light, in the middle of the tavern

16. In the Front view, move that light well below the model, as shown here:

17. In the Modify panel, set the parameters of the light:

Name the light $Below

Uncheck Cast Shadows

Set the HSV spinners to H = 0, S = 0, V = 40

Set the Multiplier to 1.0

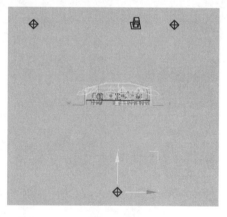

18. In the parameters of the $Below light, choose the Exclude button

19. In the Exclude / Include dialog box, highlight Ceiling in the list at left, then choose the right-pointing button to transfer Ceiling to the list at right

20. Choose the Include button at the top, then choose OK

21. Render the camera view

The illumination levels in the room look good, but the illumination on the walls and floor is too even. This scene, like every scene, needs to be "dirtied up" some. One of the easiest ways to dirty up a scene is to project a little Noise through the fill lights, to add a natural-looking variance of light across large surfaces.

Projector Light

1. Select $Fill-SW

2. In the Modify panel, choose the button labeled None in the Projector Parameters rollout

3. In the Material / Map Browser, select Noise, choose OK

4. Open the Material Editor

5. Drag the button labeled Noise from the Modify panel and drop it onto any sample window in the Material Editor, replacing the material in that window. In the Instance (Copy) dialog box, choose Instance

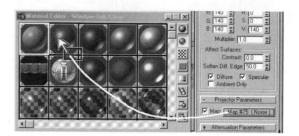

The fact that you just replaced a material in the Material Editor does not mean that material has been removed from the scene. The Material Editor can only display 24 materials at a time, and many scenes have more than 24 materials. When all the sample windows in the Material Editor are used, and you want to pull another material from a library, edit it, and apply it to the scene, just drop the new material over any sample window. The material being replaced in the Material Editor is still applied to objects in the scene, and that material can later be retrieved and dropped back into the Material Editor, if you need to edit it.

6. Set the Noise Parameters:

Set the Noise Type to Fractal

Set the Size to 60

Set the High Threshold to .9, and the Low Threshold to .1

Select the color swatch for color #1 and set that color to H = 0, S = 0, V = 125

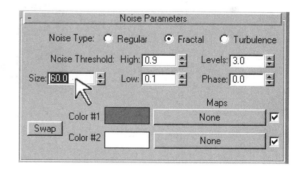

Setting the Threshold values closer together makes stronger definition of the Noise pattern. The Size of 60 means the Noise pattern will start a new iteration every 60 inches. The Fractal Noise Type takes a bit longer to render, but often looks most natural.

7. Close the Material Editor

Now that the light will be projecting this Noise pattern, the grey in the pattern will cause the light to illuminate less brightly. Increase the light's Value setting to counteract this.

8. In the Modify panel, set the light's Value to 200

Before you re-render to see the effect of the Noise, load the last rendering into RAM Player. After you render again, you can compare the two images.

9. Save the scene

Always save the scene just before using RAM Player– it may cause crashes.

10. From the menus, choose Rendering / RAM Player

11. In RAM Player, select the teapot on the left to load the last rendering into Channel A. In the RAM Player Configuration dialog box, leave all settings at the defaults and choose OK

12. Minimize (don't close) RAM Player

13. Render the camera view

14. Restore RAM Player

15. In RAM Player, select the rightmost teapot to load the last rendering into Channel B

16. Drag either of the small white triangles left and right to compare the two channels

The walls and floor now have a random variance of light that makes them more believable. Noise is a simple map to set up, but you can also create bitmaps in a paint program to project fake shadows about a room. Depending on the positions of the two fill lights in your scene, you may see in the rendering that this projector setup has created a problem– there may be seams visible on the floor or walls. In order to simplify this tutorial, Omni lights are being used as fill lights. Omnis work fine for fill lights

unless you want to project a map through them, in which case seams may appear. The two fill lights should be Free Direct lights in order to project the Noise pattern correctly. If you want to take a few extra minutes in this tutorial to remove the seams, you can correct the fill lights according to the following explanation. If you just want to know how to avoid the seams in the future, then just read the explanation.

There are two stategies for getting rid of the seams – a quick strategy that has a decent chance of working, and a slightly more involved one that will definitely work. The quick strategy is to just arbitrarily move and rotate the Omni light objects in a couple of views and try to shift the seams off any surfaces where they are obvious. With a bit of luck and a few trial renderings, this often works.

The approved method for removing the seams is to use Free Direct lights for the fill lights, instead of Omnis. A Free Direct light is either a disk or a rectangle that casts perfectly parallel rays (as opposed to the spot light, which casts a cone of light). When you are adding fill lights to a scene, and you intend to make them projectors, create the fill lights as Free Direct lights. If the fill lights are already Omni lights, there is a drop-down list at the top of the light's parameters that allows you to change the light to any other type. When you change a light from one type to another, most of the parameters are preserved. The Free Direct lights need Hotspot and Falloff values broad enough to encompass the objects in the scene (in the case of this scene, thirty feet will do), and they need to be aimed at the objects, as shown in the image at right. To be sure the lights are aimed correctly and have the correct Hotspot and Falloff values, type the $ sign to turn a viewport into a light's-eye view.

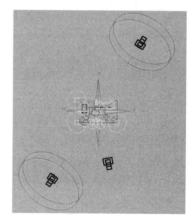

However you've left the fill lights, you have just one of them ($Fill-SW) projecting Noise. Set the other fill light to also project the Noise

17. Close RAM Player

18. Type H, select $Fill-NE

19. Open the Material Editor

20. In the Modify panel, open the Projector Parameters rollout, then drag the Noise sample window from the Material Editor and drop it onto the button labeled None in the Projector Parameters rollout

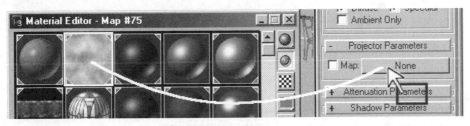

21. In the Instance (Copy) Map dialog box, choose Instance

22. Set the light's Value to 200 to counteract the darkening effect of the Noise map

23. Close the Material Editor, then save the scene

Shadow-Only Light

There is one last lighting issue to resolve for this scene. There should be shadows from the furnishings, appearing to be cast by various light around the room. This is easily accomplished with a couple lights that will cast no illumination, but still produce shadows.

1. Make the Top view active

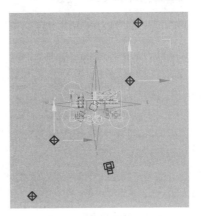

2. Use Shift-Move to duplicate both fill lights, setting the two duplicates closer to the building, as shown here:

These need to be Omni lights. If you converted the fill lights to Free Direct lights, convert these duplicates to Omnis.

3. Name the new lights $Shadow-SW, and $Shadow-NE

4. In the Modify panel, set each light's parameters:

> Make sure Cast Shadows is checked
>
> Set the Multiplier to 0

In the Projector Parameters rollout:

> Uncheck the Map check box to disable the Noise

In the Shadow Parameters rollout:

> Set the shadow Color to pure white:
> H = 0 S = 0 V = 255
>
> Set the shadow Density to -.5 (note the minus)

In the Shadow Map Params rollout:

> Set map Size to 512 and Sample Range to 8
>
> Check Absolute Map Bias

The online help offers this in the way of explanation of Absolute Map Bias: "When on, the bias for the shadow map is computed on an absolute basis, relative to all of the other objects in the design. When it's off, the bias is computed relative to the rest of the design". Not too helpful an explanation– in fact, it seems redundant. Absolute Map Bias seems to use scene units for bias values instead of arbitrary units. If your shadows are not touching their objects, and setting Bias to .1 with Absolute Map Bias unchecked does not solve the problem, try checking it.

> Choose the Exclude button
>
> In the Exclude / Include dialog box, transfer Bar, Beer Taps, Floor, [Bar Stools], [Chairs], [Pool Table], and [Tables] from the list at left to the list at right
>
> Select the Include button, then choose OK

Repeat parameters setup for the other shadow-only light

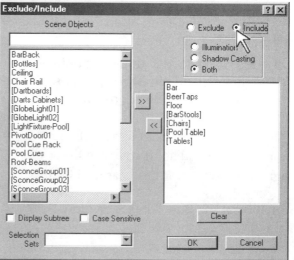

5. Render the camera view

The settings for a shadow-only light are not found in the books that ship with the program, or in the online Help. This is a trick submitted by a visitor to the online VIZ support forum at www.vizonline.com. This forum is really the only reliable source of technical support available to all users of VIZ. If you set up your e-mail program to monitor the forum, you will see new threads and postings each time you check your e-mail, and you can reply to postings through e-mail.

6. Save the scene

Final Rendering - Panoramic

This is a good scene for trying out the Smoothmove 360-degree renderer, which renders images in six directions from one point of view, seams the six renderings together, maps the panoramic bitmap onto the inside of a spherical environment, and allows the viewer to move the mouse to freely look in any direction in that environment.

1. From the menus choose Rendering / Environment

2. At the top-left of the dialog box is the color swatch showing the blue-gray you've been seeing through the windows in renderings. Next to it is a button labeled Facade7.jpg. Check the Use Map check box above that button to activate the environment map.

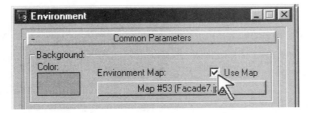

Back on page 311 you turned off shadow-casting for the two globe lights, to save rendering time. You will want those shadows for the final rendering.

353

3. Type H, select $Globe01

4. In the Modify panel, check the Cast Shadows check box

Because $Globe02 is an instance of $Globe01, it too has had shadow-casting turned on.

5. From the menus, choose Rendering / Effects

6. In the Rendering Effects dialog box, highlight Lens Effects, then highlight Glow-Neon Sign in the list of scene effects, then in the Glow Element rollout (not the Lens Effects Globals rollout), set the Size to .5

Remember that the size of a lens effect is a percentage of rendered image size, so the size that works when the camera is close to the object creating the effect will be too large when the camera is further away.

7. Make the camera view active, right-click over the view's label in the upper-left of the view, and from the pop-up menu choose Views / Camera02-Panorama

8. Save the scene

9. Switch to the Utilities panel, choose Smoothmove Panoramas SE, choose Render

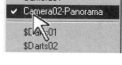

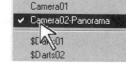

354

Rendering a panoramic view takes a while, maybe 15 minutes, so if you have other work to do at this point, you might want to come back to this rendering at the end of your day.

10. In the Output Size field, select the preset button labeled 3072 x 1536

11. In the Render Output field, check the Save File check box

12. Select the Pan Files button. In the browser, navigate to C:\Viztutorials\Chapter6, name the file Tavern01.pan, and choose Save

13. Check JPEG Compression, and set the Quality spinner to 90

14. Select Render

After rendering views looking Front, Back, Left, Right, Up, Down, and seaming those six views into one image, the Smoothmove renderer will open the .pan file for viewing. Size the view larger if you want. To view the room, just drag anywhere over the image. To zoom in and out, hold the Ctrl key on the keyboard as you drag. To set the viewer back to the initial view, press the spacebar.

This version of Smoothmove that ships with VIZ is the "light" version, in which you can look about freely, but only from one point of view. You can purchase the full package, which allows you to define a path through the model, and event hotspots along the path, so that the final product is a truly interactive experience. You can move along the path and look all about, and you can touch hotspots to activate Web pages, play animations, or jump to other Smoothmove files.

Uploading .pan files for your client to download and view is an amazing way to communicate. If you intend to send them over the Web, use the .jpg compression when setting up the rendering. Uncompressed .pan files are huge. The client will need to also download the free Smoothmove viewer to view the .pan file. The viewer is available at www.smoothmove.com. Follow the link for IMove Free Viewer, scroll to the very bottom of that page, and you will find the choice to download a version of the viewer that plugs into your Web browser, or a stand-alone version.

⚠ Once you close the .pan file that you are currently viewing in VIZ, you won't be able to open it in VIZ again. You must download and install the viewer to view .pan files that you have saved to your hard drive.

Summary

Hopefully you have noticed as you've worked through this chapter that the materials in this scene are extremely simple. Only three bitmaps are used in the scene: the dartboard image, the stained-glass skylight image, and the environment map you see through the windows. The materials were left as simple colors to keep your focus on lighting, and to underscore the importance of light and shadow. Considering that no attempt was made to develop realistic materials, it's a pretty good scene. While the image is not photo-realistic, the renderings are attractive and convey the mood of the room.

When you begin to run out of time on a project, and you realize you do not have enough time to develop both a great lighting scheme and great materials, focus on the lighting. If your client will allow it and it is appropriate to the scene, take some liberties with the real-life lighting situation for the space, to get lighting features that will add drama to the images. Remind your client that the ultimate purpose of a rendering is to sell the design idea, and in selling, image is often more important than truth.

CHAPTER 7

Man-Made Materials

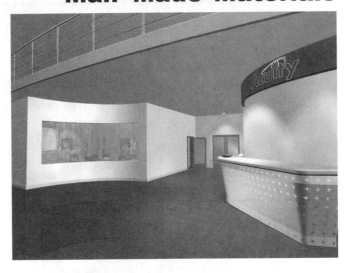

The Material Editor is probably the most complex area of VIZ, and it is also the area in which your creativity is most important. Learning the settings is only half the challenge. In time you will gain the experience and intuition that will allow you to use the settings in novel ways to achieve any effect you need. Learning to develop materials is like learning to paint. It requires that you reexamine all the mundane items that surround you each day, recognizing and suppressing assumptions that your mind's eye has been making about these objects all your life and seeing things for what they really are. It may sound a bit lofty to say, but you will actually see the world a bit differently once you've spent hundreds of hours interpreting its qualities on a computer.

In the prebuilt scene for which you will develop materials, the lighting scheme is already in place. You would not work this way in reality. Lighting and materials are not so much two subjects as they are two parts of the same subject. You would be unwise to spend hours on a lighting scheme without deciding the qualities of the surfaces upon which the light will fall. And you cannot develop a good palette of materials in an environment of generic, shadowless lighting. When developing materials for an actual project, force yourself to leave the Material Editor occasionally and adjust the qualities and positions of the lights. You'll be surprised at the great difference that a small change in the intensity or the angle of a light makes in the appearance of the materials.

Open the Scene

1. From the menus choose File / Open,
   browse to C:\Viztutorials\Chapter7,
   and open Lobby01.max

2. Quick Render the camera view

The view outside is a high-resolution image
taken with a digital camera. In Photoshop,
a new sky and asphalt texture were added.

Shader Types / Basic Parameters

The first material to build is for the columns supporting the portico roof.
Initially you will make these look like basic silver metal, and a bit later, in
the discussion of various map types, you'll add a raised pattern of bumps
to the metal. Currently the columns, like most of the objects in the
scene, carry a neutral light gray material.

1. Resize and scroll the rendered view so that the column is visible in
   a narrow window

2. Open the Material Editor. If there are not fifteen sample
   windows showing, right-click over the active sample window and
   choose 5 x 3 Sample Windows

3. Make the fourth
   sample window active

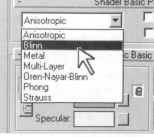

4. Open the Shader type drop-down list, view the
   choices, then set it to Blinn

Shaders are mathematics that describe how light interacts
with a surface. Several of these shaders are named for the
mathematicians who developed them. There are seven to
choose from, but in general practice you will likely use
Blinn, Anisotropic, Metal, and Oren-Nayar Blinn nearly all the time.

Blinn is the most commonly used shader. It is appropriate for wood, plastic, concrete, earth, fabric, and gypsum; basically anything except polished, hard, curved surfaces.

The **Anisotropic** shader allows for specular highlights that are not round. You might use it to depict a brushed aluminum surface. It is usually the best choice for flat sheet metal.

Metal is for curved metal, and often works well for curved glass.

Oren-Nayar Blinn specializes in depicting matte fabric, like felt or carpet.

Multi-Layer is two layers of Anisotropic.

Phong shading was one of the earliest shaders written. Its function is largely replaced in VIZ3 by Blinn.

Strauss is another (less effective, it seems) shader for curved metal.

The qualities that these various shaders define pertain more to curved surfaces than to flat ones (which presents the first challenge when designing materials for architectural models, where flat surfaces are much more common than curved ones). Essentially what a shader defines is the quality of a surface's Specular highlight, Diffuse area, and Ambient area.

The Diffuse area of an object is the area of general illumination.

The shiny spots are the Specular highlights.

The part in shadow is called the Ambient area.

The size, intensity, and shape of specular highlights are a strong visual signal about the quality of a surface. When we see the small, intense highlight on a ball bearing we expect that surface to be cool, smooth, and hard. When we see the broad, soft highlights on a leather sofa, we sense the give of the cushions and the grain of the skin. Your sense memory would be confused by the appearance of a highlight on the felt surface of a pool table. Your job in a visualization is to trigger visual and sense memories in the viewer, to convey the feeling of being in a space. The great challenge is that so many of our visual and sense memories are sparked by subtle clues of which we are not even aware. Be observant of the everyday, and try to detect and remember those subtle signals.

Blinn Shader

5. Name the material Columns

6. Drag the material from the sample window and drop it onto the column in the camera view. To ensure that the material is being dropped on the correct object, pause for a moment before releasing, and wait for the tooltip to identify the object

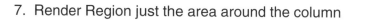

7. Render Region just the area around the column

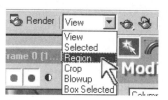

Currently the column looks like it might be made of vinyl. By focusing the specular highlight and making it more intense you can make it look like hard, polished plastic.

8. Set the Specular Level spinner to 80, and the Glossiness spinner to 30

9. Choose Render Last (which will render the same region as before)

The column now looks very different, primarily because of the specular highlight. The color and intensity of the diffuse area has not changed. The ambient area is a bit darker.

Specular Level is the intensity, or brightness of the specular highlight.

Glossiness is the diameter of the specular highlight. The higher the setting, the smaller the diameter, and the shinier the object.

Notice the graphic representation of these two settings, to the right of them. You will learn to recognize certain shapes as producing certain looks, and you may come to rely as much on the appearance of the curve as you do on the numbers in the spinners to get the look you want.

Dull plastic-
Blinn shader  Often works
well for walls,
floors, ceilings Polished metal,
glass (curved).
Metal shader

Colors: Ambient, Diffuse, Specular

When you are asked the color of an object, you will likely reply with one color: "the sofa is green." In VIZ, only the Strauss shader uses a one-color definition. The rest ask you to specify at least two colors, and sometimes as many as four. There might be a color swatch for the area of the object in shadow (Ambient), the area of general illumination (Diffuse), the color of the Specular highlight, and if the object has transparency, the color of light traveling through the object (called the Filter color).

Make the column lighter in color.

1. Choose the Diffuse color swatch, and set the diffuse color to: H = 140, S = 10, V = 240

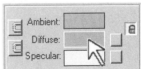

2. Render Last

In most cases, it will work well to set the diffuse and the ambient colors to the same hue, and set the value of the ambient color to a lower value. A greater sense of depth can be achieved by setting the ambient color to a very dark and saturated color. Set the ambient area of the column to a dark purple.

361

1. The Ambient and Diffuse color swatches are locked together. Unlock them with the U-lock button to their left

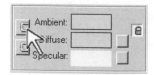

2. Choose the Ambient color swatch and set it to:
   H = 195, S = 200, V = 70

3. Render Last

The right side of the column darkens, and while the purple cast it takes on is not obvious enough that a viewer would notice it as being purple, it is there if you look for it.

4. Set the Ambient back to the same color as the Diffuse by relocking them

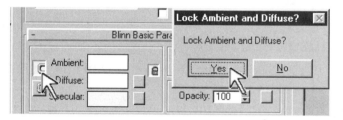

The Specular color should be set to the color of the light that is creating the specular highlights on the object. For normal interior lighting this can be left at white. For an object in bright sun, tint it a pale yellow. If the object is under a colored feature light, tint the specular color. Depending on the material, the specular color that gives the effect you want may or may not be a paler version of the color of the feature light.

Metal Shader

To make the column look more like metal, there should be more distinct specular high-lights, and greater contrast between the specular, diffuse, and ambient areas. The Metal shader accomplishes this.

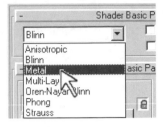

1. Set the shader type drop-down list to Metal

Notice there is no Specular color swatch in the Metal shader. The color of the specular highlight will be derived from the color of the diffuse area. You do not have an opportunity to set that color yourself.

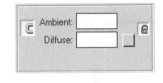

2. Set the Specular Level to 45, and the Glossiness to 75

3. Render Last

Not only are the highlights more distinct, there are more of them (look at the one low on the right side).

The Specular Level and Glossiness settings work differently in the Metal shader than they do in the Blinn shader (and admittedly, the way they work is a bit confusing and not easy to remember).

Remember that in Blinn shading, Specular Level is the intensity (or whiteness) of the highlight, and Glossiness is the diameter of the highlight- the smaller the diameter, the shinier the object appears. In Metal shading,

Glossiness does both jobs; the higher the Glossiness, the smaller and more intense (white) the highlight. To see this work, open a larger sample window for this metal column material and change the settings as follows.

4. Double-click the sample sphere for the column material

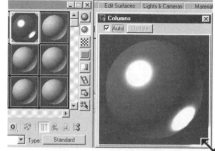

A larger sample window appears

5. Size the new floating sample window larger (about twice its present size), by dragging a corner of it

6. Leave the Specular Level at 45, set the Glossiness to 50

The specular highlight gets very broad, and less white.

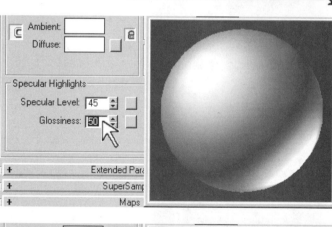

7. Leave the Specular Level at 45, set the Glossiness to 95

The specular highlight gets narrow, and very white

So if, in Metal, the Glossiness setting has taken over both duties performed by the two settings in Blinn, then what

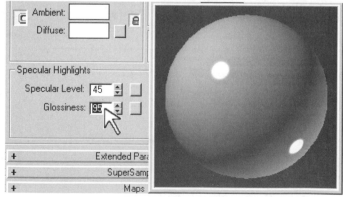

does Specular Level do in Metal? It doesn't affect the specular highlight, it affects the rest of the object.

8. Leave the Glossiness at 95, set the Specular Level to 30

The highlight is unaffected and the rest of the object brightens.

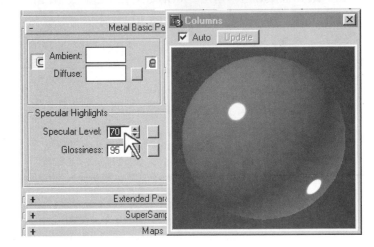

9. Set the Specular Level to 70

The highlight is unaffected, and the rest of the object darkens.

Why was this set up this way? Only the programmers know for sure, but here is a hypothesis. If this sphere were a highly polished, large ball bearing, and it was held under a very bright light, the specular highlight would be so intense it would hurt the eyes. A computer cannot emit light this bright, and neither can paper. Once the highlight is at pure white, that is as intense a depiction as a monitor or a piece of paper can achieve. The only way to make the highlight appear more intense is to darken the rest of the object, so that the specular highlight stands in greater contrast to the diffuse area, and so seems brighter.

10. Set the Specular Level to 45, and the Glossiness to 75

11. Close the larger floating sample window

12. From the menus choose File / Save As. Navigate to the C:\Viztutorials\Chapter7 folder, name the scene Lobbyxx.max, where xx is your initials, and choose Save

Anisotropic Shader

The dictionary definition of anisotropic is "exhibiting properties with different values when measured along axes in different directions." Applied to this shader, it means that the specular highlight on the object can be elongated into an ellipse and oriented in any desired direction. The most useful application of anisotropy in VIZ is probably in the depicting of flat metal. The anisotropic shader is new to VIZ 3, and before it was added, flat metal was arguably the most difficult thing to depict convincingly in VIZ. As stated before, the properties of specular level and glossiness pertain more to curved objects than to flat ones. A sheet of metal does not have small, round, intense specular highlights to tell you that it is metal (unless it is highly polished and there are lights in just the right place). Rather it has a broad band of highlighting, or more of a gradient than what is usually thought of as a highlight. To make such a band appear on a flat surface you need two things: an anisotropic material applied to the surface, and a strategically placed light to make the highlight happen in the right place.

1. Drag the Time Slider to frame 40

2. With the camera view active, set the render type drop-down list to Blowup

Blowup renders a magnified portion of the view.

3. Choose the leftmost Render teapot to open the Render Design dialog box.

4. In the dialog box, select the 320 x 240 output size button

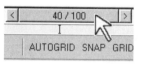

5. Move the dialog box so you can see the camera view, and choose the Render button at the bottom of the dialog box

6. A dashed rectangle appears in the camera view. Move and resize it so it is the height of the elevator doors, and centered on the doors, as shown at right:

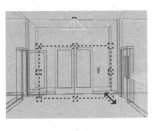

7. Choose the OK button in the lower-right of the view

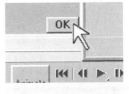

The doors have a Multi/Sub-Object material applied to them. In the mesh, the frames have been assigned material ID 1, and the panel inserts have been assigned material ID 2. You'll load the door material into the Material Editor, change the color of the frame submaterial, and set the anisotropic properties of the panel material.

8. In the Material Editor, make the fifth sample window active

9. Choose the eyedropper button, (Pick Material from Object) to the left of the material name

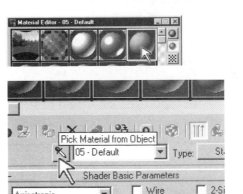

10. In the camera view, place the eyedropper over the elevator doors, pause a moment to see the tooltip appear to verify that you are on the Doors-Elevator object, then click to load the door material into the Material Editor

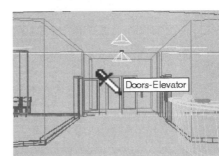

367

11. Choose the button labeled Frame (Standard) to drop to that sub-material

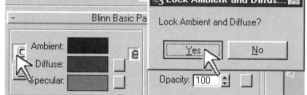

12. Lock the Ambient and Diffuse color swatches

13. Choose either color swatch and set the colors to H = 170, S = 15, V = 170

14. Choose the Go Forward to Sibling button to switch to the other sub-material (the Panels material)

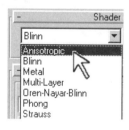

15. Set the shader type drop-down list to Anisotropic

16. Lock the Ambient and Diffuse color swatches, then set them to:
    H = 138, S = 10, V = 200

17. Set the Specular color swatch to: H = 138, S = 10, V = 255

18. Set the Specular Level to 70

19. Set the Glossiness to 75

20. Set the Anisotropy to 80

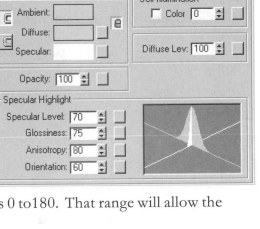

Anisotropy of 80 will produce a very elongated highlight.

21. Set the Orientation to 60

The practical range for the Orientation spinner is 0 to180. That range will allow the highlight to rotate to any angle you want.

22. Render Last

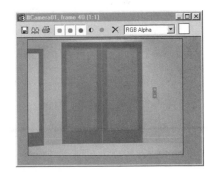

The specular highlights are not happening yet. A well-placed light is needed. This light will be set to only illuminate the elevator doors, and it will also be set to only cast specular illumination; it will not brighten the diffuse or ambient areas of the doors.

Place Highlight

1. Close the Material Editor

2. Make the Top view active

3. In the Create panel, Lights category, choose Omni

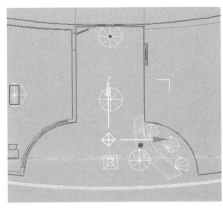

4. Create the omni light in front of the reception desk, near the start of the hall leading to the elevator, as shown at right:

5. Make the camera view active

6. At the bottom of the VIZ interface, lock the selection

Locking the light selection isn't a requirement in the Place Highlight tool, but the light often becomes accidentally unselected during the next step, so it helps to lock the selection to prevent that.

7. From the toolbar at the left of the views, choose the Place Highlight tool

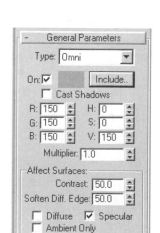

8. Position the cursor anywhere over the elevator doors, click-hold, and drag across the surfaces of the doors. The omni light will move to align itself so that its rays strike the point where the cursor is, and bounce into the camera. Move the cursor to one of the frame elements at the middle of the doors, and not quite halfway up the door, as shown here. Release the mouse button

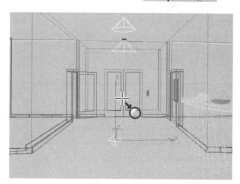

9. Unlock the Lock Selection Set button

10. In the Modify panel, set the light's parameters:

Name the light $ElevatorHighlight

Uncheck Cast Shadows

Choose the Exclude button, and in the Exclude / Include dialog box, transfer Doors-Elevator to the list at right, choose the Include button, choose OK

Set the HSV spinners to H = 0, S = 0, V = 150

Set the Contrast spinner to 50

Uncheck the Diffuse check box so that the light will only cast specular illumination

370

11. Render Last

12. Save the scene

Getting the highlight-casting omni light in just the right place usually takes a bit of trial and error, but it shouldn't take long to place a few in a scene to get the flat metal surfaces to shine correctly. Because the light is set to only include certain objects, and to not cast shadows, it will not increase rendering times noticably to add several of these highlight-casters to the scene.

Oren-Nayar Blinn Shader

This shader is good for very matte surfaces, and the green carpet in the conference room is a perfect opportunity to examine what it does. The best way to see its effect is to first render the carpet using a Blinn-shaded material, then using an Oren-Nayar Blinn-shaded material, and compare the two in RAM Player.

1. Right-click over the viewport label of the camera view (the word #Camera01), and from the pop-up menu choose Views / #Camera02

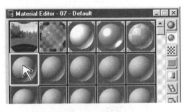

2. Type H, choose Carpet

3. In the Material Editor, make the leftmost sample window in the second row active

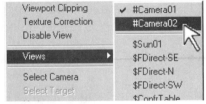

4. Name the material Green Carpet

5. Set the shader type drop-down list to Blinn

6. Lock the Specular and Diffuse colors (the Ambient and Diffuse should already be locked)

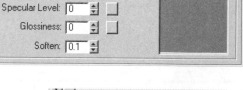

7. Set the colors to H = 103, S = 45, V = 108

8. Set the Specular Level to 0, set the Glossiness to 0

You are making the material as matte as Blinn shading will allow.

9. Choose the Assign Material to Selection button

10. Set the render type drop-down list to View

11. Choose the leftmost Render teapot, and in the Render Design dialog box choose the 640 x 480 preset output size button, then choose Render

12. Quick Render the camera view

Even with the shader settings set to completely matte, the specular area of the carpet is still too distinct at its edges, and too bright at the center.

Load this image into RAM Player for comparison to Oren-Nayar Blinn.

13. Save the scene

14. From the menus choose Rendering / RAM Player

15. Choose the left teapot, Open Last Rendered Image in Channel A. In the RAM Player Configuration dialog box, leave all settings at the defaults, and choose OK

16. Minimize RAM Player (don't close it)

17. In the Material Editor, in the Green Carpet material, set the shader type drop-down list to Oren-Nayar Blinn

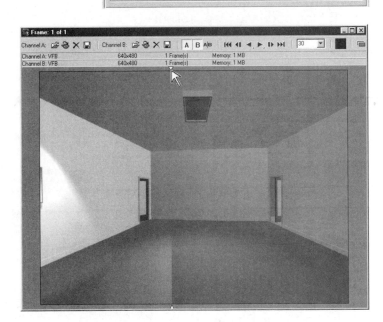

18. In the Advanced Diffuse group, set the Roughness to 100

A rough surface, like carpet or terra cotta, scatters light and doesn't allow distinct specular areas.

19. Render Last

20. Restore RAM Player, choose the right-hand teapot to load the last rendering into channel B. Drag either small white triangle to compare the two channels

The Oren-Nayer Blinn shader produces a much softer and more even specular area, which will make a more believable carpet.

21. Save the scene

22. Close RAM Player

In the next section you will replace the basic green color with a scanned photograph of green carpet to make the material more realistic.

Diffuse Color Map

For objects in the distant background of an image, conveying the right color and the right specular qualities is often enough to make the object realistic. For closer objects, you usually need more than just a base color; a map of some sort needs to be incorporated into the material. The map might be a bitmap; maybe an image scanned from a magazine or a fabric sample, or taken with a digital camera. Or it might be a procedural map, not saved on the hard drive but generated entirely mathematically by VIZ at rendering time. A map that imparts a pattern over the entire surface is called a Diffuse Color map. Maps are also incorporated into materials to serve more specialized functions: a Bump map to simulate bumps, an Opacity map to make parts of the surface disappear, a Specular Level map to control where the surface shines brightly, and where it is matte, or a Reflection map to suggest reflections without having to calculate the true ones.

Replace the basic green color of the carpet with a scanned photograph of carpet.

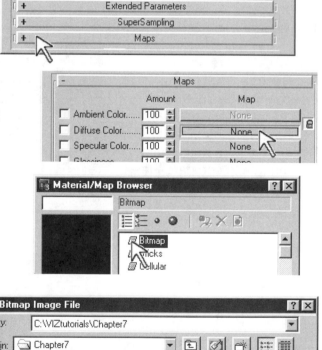

1. At the bottom of the Material Editor, open the Maps rollout

2. Choose the wide button labeled None in the Diffuse Color channel

3. In the Material / Map Browser, highlight Bitmap, and choose OK

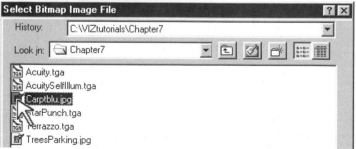

4. In the Select Bitmap Image File dialog box, browse to C:\Viztutorials\Chapter7, highlight Carptblu.jpg, and choose Open

374

When you chose Open, you dropped to a lower level of the Material Editor. It is not uncommon to have three or four levels involved in a complex material, and each level may have more than one sibling set of parameters, each on its own screen. Unfortunately, the various levels and sibling parameter screens aren't visually differentiated. The key to knowing where you are as you navigate and adjust the various levels and siblings of a complex material is to give logical names to each level and sibling. This carpet material will be relatively simple. Only a Diffuse Color map will be incorporated into the material so this is the only sublevel screen that needs a name.

5. Name this level of the material Carpet Bitmap

Notice, to the left of the name, the label Diffuse Color, identifying the function of the current level or sibling. A bit later, you'll use the Material / Map Navigator, which is a floating window showing all the level and sibling labels arranged in a hierarchical tree.

6. In the Bitmap Parameters rollout, choose the View Image button

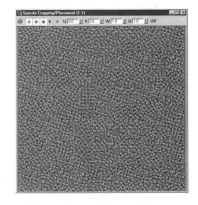

This image was most likely acquired by laying a carpet sample on a scanner. It probably represents one square foot of carpet. When you use a Diffuse Color map of a real-world material, you need to view the map and decide what the real-world dimensions of the map are. For example, if the bitmap shows six courses of brick, you would find in a reference book that six courses of standard brick is approximately 1′4″, so this is the vertical real-world dimension of the bitmap. This information will be needed to apply proper mapping coordinates to the object receiving the material. Mapping coordinates are instructions for the scale and orientation of maps applied to an object. Mapping coordinates were described in Chapter 3 (page 143), and will be discussed again in this chapter.

Tinting a Bitmap

The most obvious problem with this carpet image is that it is blue, and the carpet in the conference room should be green. Fortunately, altering the colors of a bitmap is easy.

1. Close the window displaying the carpet bitmap

2. At the bottom of the Material Editor, open the Output rollout

3. Put a check in the Enable Color Map check box

4. Change the color map from Mono to RGB (red, green, blue)

5. Choose the R button and the G button to deactivate them, leaving only the B button active

6. Position the cursor over the vertex at the upper-right end of the color map graph, and drag downwards. As you drag, a numeric field below the graph window updates. Drag the vertex so that the numeric field reads approximately .7, and release. If needed, fine-tune the setting by typing .7 in the numeric field

7. Render Region, to render just the carpet area

The color and the specular qualities of the carpet are good. The texture, however, is uneven and shows bands of odd patterning in places. This is often due to flaws in the scanned image being applied, but it is also commonly due to the need for SuperSampling. Add SuperSampling to improve the depiction of the carpet bitmap.

376

SuperSampling

A rendered image is made up of a certain number of pixels across, and a certain number of pixels down (for instance 640 x 480 or 800 x 600). Each pixel represents, or covers, a certain real-world distance in the scene. If this image of the conference room is rendered at 800 x 600, and the distance across the conference room floor near the camera is roughly 230 inches, then each pixel of the rendered image represents approximately .29 inches across, (230/800) or just over a quarter-inch of carpet. There are several colors present in a quarter-inch of carpet. A rendered pixel can only be one color. How does the renderer decide which of the several carpet colors in a quarter-inch to assign to a particular pixel? Without SuperSampling, the renderer uses the color at the center of the area that a pixel covers to decide the color of the pixel. This method often results in odd patterns or poorly defined edges. SuperSampling examines nearby pixels and uses one of several averaging methods to make a more informed decision about the correct color for each pixel. This extra calculation takes considerable additional time, but often makes a huge difference in the quality of the depiction of a bitmap on a surface.

1. Choose the Go to Parent button to return to the topmost level of the carpet material

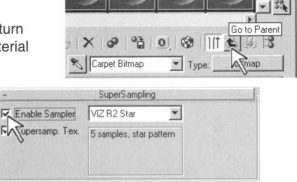

2. Open the SuperSampling rollout

3. Check Enable Sampler

4. Save the scene

5. Render Last

This is a much better carpet pattern, at a cost of greatly increased rendering time. In the Render Design dialog box is a check box labeled Disable all Samplers. You could use it to speed rendering while you develop materials and lighting, then enable all samplers for the final rendering.

Self-Illumination Map

Self-illumination makes an object appear to glow. The way it does this is simple; it brightens the object's ambient areas. When the parts of an object that are in shadow exhibit a uniform brightness, we understand that as a glowing object. Assigning self-illumination can be as easy as setting a single spinner value, but such an application of totally uniform self-illumination over an object rarely looks realistic. Self-illumination is usually applied using a grayscale map. Lighter areas of the map make the object more self-illuminated, darker areas less so. Examine self-illumination by designing a material for the diffuser in the light fixture in the conference room.

1. Type H, select the Diffuser

2. In the Material Editor, make the next unused sample window active

3. Name the material Diffuser

4. Set the shader type drop-down list to Blinn

5. Lock the Diffuse and Specular colors together (the Ambient and Diffuse should already be locked), and set the colors to H = 141, S = 15, V = 240

6. Set the Specular Level to 80 and the Glossiness to 10

7. Set the Self-Illumination to 80

8. Choose the Assign Material to Selection button

378

9. Set the render type drop-down list to Render Blowup

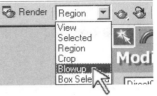

10. Choose the leftmost Render teapot to show the Render Design dialog box. Choose the 320 x 240 output size, and choose Render

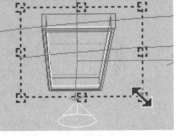

11. In the camera view, size and move the blowup rectangle so it just fits around the light fixture, then choose the OK button in the lower-right of the camera view

The diffuser is glowing, but it certainly doesn't look realistic. It needs to show the texture of the plastic diffuser, and the way this texture mitigates the glow of the light. Add a bitmap to the Self-Illumination channel to achieve this effect.

12. In the Maps rollout (open it if it is closed), choose the button labeled None in the Self-Illumination channel

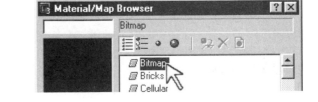

13. In the Material / Map Browser, highlight Bitmap, and choose OK

14. In the Select Bitmap Image File dialog box, browse to C:\Viztutorials\Chapter7, highlight Diffuser.tga, and choose Open

15. Render Last

The bitmap is doing the job, but it's overstated. Lower the percent at which the bitmap is used.

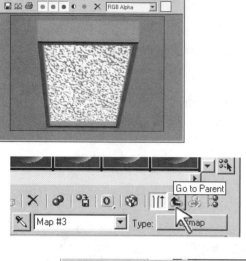

16. Choose the Go to Parent button

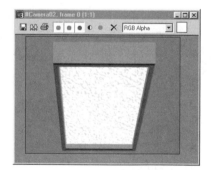

17. Set the spinner for percent usage to 30

18. Render Last

When you use a map channel at less than 100 percent, the corresponding setting in the Basic Parameters rollout for that characteristic of the material mixes in with the map channel to add up to 100 percent. For example, if the Diffuse color in the Basic Parameters rollout is red, and you assign a bitmap in the Diffuse Color channel that is blue, and set that channel to be used at 60 percent, the result will be a purple material; 60 percent blue map, 40 percent red base color.

19. Save the scene

What this diffuser needs to look more convincing is some Glow effect, such as what was used for the neon tubes in Chapter 6 (page 337). If you're feeling ambitious, try setting this up. Start by choosing, from the menus, Modify / Properties, and setting the G-buffer Object Channel to 1. Then choose Rendering / Effects / Add, and add a Lens Effect. Select Glow as the desired lens effect. Leave the settings in the Lens Effects Globals rollout at the defaults. In the Glow Element rollout, try the Size at 5, the Intensity at 60, and choose a bright and slightly blue color for the Radial Color. In the Options tab, Image Sources group, check the box for Object ID 1, and render.

Specular Level Map

A Specular Level map is generally a grayscale map, in which lighter areas of the map produce brighter areas on an object's surface. One of the most common uses of a Specular Level map is to add a natural, random variance of brightness to a large surface (in the previous chapter you projected Noise through the fill lights to accomplish this). Another example of Specular Level mapping is in a tile floor or a tiled bathroom. The tiles and the grout lines are part of the same material (you'll make such a material shortly). The tiles should have a sheen to them, while the grout lines should be completely matte. A Specular Level map showing a white square with a thin black border would make shiny tiles (the white square) with dull grout lines (black border).

Use a Specular Level map to vary the brightness of the walls and ceilings.

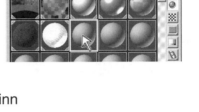

1. In the Material Editor, make the next unused sample window active

2. Name the material Walls-White

3. Set the shader type drop-down list to Blinn

4. Set the Ambient and Diffuse colors to H = 30, S = 15, V = 240

5. Set the Specular Level to 35, and the Glossiness to 5

6. Open the Maps rollout if it is closed

7. Choose the button labeled None in the Specular Level channel

8. In the Material / Map Browser, highlight Water, and choose OK

Water on walls? Water is a procedural map; the pattern is generated mathematically at rendering. The names given to procedural maps are merely suggestive. Water works well for conveying the variance of light across walls because the pattern has layers of overlapping diagonal shading that suggest ambient light bouncing from the architectural elements of the scene.

Water Procedural Map

To understand the settings for the Water procedural map, imagine small stones being dropped from several locations above, onto the surface of a still pond. When the rings from the various stones all meet and interfere, the result is the Water map. The Water settings address the size of the stones, the distance between stones, the height of the waves, and the colors of the crests and troughs of the waves.

9. In the Coordinates rollout, set all three Tiling spinners to .45

The smaller the Tiling values, the larger the scale of the pattern.

10. In the Water Parameters rollout, set:

Num Wave Sets to 10

Wave Radius to 300

Wave Length Max to 25

Wave Length Min to 1

Color #1 to totally black

Color #2 to totally white

Num Wave Sets is the number of stones being dropped on the pond. Wave Radius is the distance between stones; as the rings travel away from the point of impact, their radius increases. Larger radius values produce a more linear pattern of diagonals. With a smaller radius you can see the arcs of the rings. Wave Length Max and Wave Length Min refer to the sizes of the stones, and the variance in size among the stones. Bigger stones make bigger wavelengths. What is important here is the difference in size. Max and Min settings far apart (large variance in stone sizes) produce more chaotic patterns, while Max and Min settings close together produce more distinct patterns. Amplitude is the height of the waves. When you use Water to actually make a water surface, Color #1 will be the color of the crests of waves, and Color #2 will be the color of the troughs. Used as a Specular Level map, the dark areas will be less bright, the light areas more bright. Note that the color values used here are the opposite of what they would be for water– these colors would make dark crests and bright troughs. Swapping the values seems to work better for walls.

11. Drag the sample sphere for the wall material and drop it onto any wall in the active view. Before releasing over a wall, hold the mouse still long enough to see the tooltip appear identifying the object under the cursor. You won't be able to drop onto the middle of a wall plane. You need to find an edge before the cursor shows the material drag icon

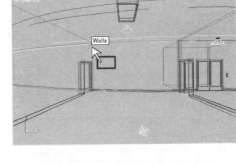

12. Right-click over the view label of the camera view and choose Views / #Camera01

13. Drag the Time Slider to frame 28

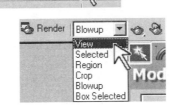

14. Set the render type drop-down list to View, then choose the leftmost teapot

Draft Render

When your scene uses Lens Effects, Volume lights, and SuperSampling, and you want to do a fast render to test something that doesn't involve these things, you can set up Draft render settings and Production (final) render settings.

15. In the Render Design dialog box, choose the Draft button at the lower-left of the dialog box

16. In the Output Size group, choose the 640 x 480 preset size button

17. In the Options group, uncheck all check boxes (Atmospherics, Effects, etc.).

18. In the Global SuperSampling group, check Disable all Samplers

19. In the VIZ Default Scanline A-Buffer rollout, put checks in the Mapping and Shadows check boxes

20. Choose Render

Notice that after rendering a Draft render, the Quick Render button shows a gray teapot. That button is a flyout, with choices for Production, Draft, and Hidden-Line render.

The Specular Level effect is much too pronounced. Lower the percent at which the Water map is used in the Specular Level channel to make the effect more subtle.

20. In the Material Editor, choose the Go to Parent button

21. Set the percent usage spinner in the Specular Level channel to 40

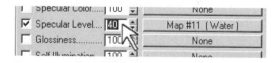

The Water map is now contributing 40 percent of the specular level quality of the walls. The other 60 percent is being controlled by the Specular Level setting in the Blinn Basic Parameters, which is set to 35.

22. Save the scene

Bump Map

Bump mapping, like many of the maps channels, generally uses a grayscale map, in which the light parts of the map give the appearance of a raised area of the surface, and the dark parts give the appearance of a depression in the surface.

Return to the Columns material from the beginning of the chapter, and add a pattern of raised bumps to the metal column covers.

1. Drag the Time Slider to frame 0

2. In the Material Editor, make the Columns material the active sample window

3. In the Maps channels, choose the button labeled None in the Bump channel

4. In the Material / Map Browser, highlight Bitmap, and choose OK

5. In the Select Bitmap Image File browser, browse to C:\Viztutorials\Chapter7, highlight Dot01.tga, and choose Open

6. In the Coordinates rollout, set the Tiling U: spinner and the Tiling V: spinner both to 12

Tiling is how many times the image repeats over a given distance. This distance is given by the UVW Map modifier, which has already been applied to the columns.

385

7. Quick Render the camera view, still using Draft render settings

The bumps are visible, but too faint. Increase the strength at which the bump map is used.

8. Choose the Go to Parent button

9. In the Maps channels, set the Bump map strength spinner to 125

10. Render Last

Most of the usage strength spinners in the maps channels have a maximum value of 100, but the Specular Level, Bump, and Displacement spinners have a maximum value of 999.

Have a look at the UVW Map modifier applied to the columns.

11. Type H, select Columns

12. In the Modify panel, turn on Sub-Object

In the camera view you can see the mapping Gizmo highlighted yellow. In the Modify panel are the parameters for the UVW Map modifier. It is set to type Cylindrical, which makes sense for a column, and the dimensions of the Gizmo are set to:

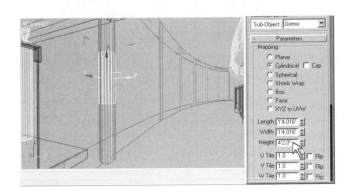

> Length = 1'4"
> Width = 1'4"
> Height = 4'2"

When the UVW Map modifier was applied, its Length and Width automatically sized themselves to fit the dimensions of the column, as did the height. The height was adjusted to be the same as the circumference of the column. The radius of the column is 8", so the circumference is approximately 50", or 4'2" (circumference = 2pi x radius, or 6.28 x 8").

In the Material Editor, in the Coordinates rollout for the Bump map, you set the Tiling for both the U and the V directions to 12, thereby instructing the Dot01.tga to repeat itself twelve times around the Gizmo, and twelve times up it. Setting the height of the Gizmo to equal its circumference ensures that when the same value is entered in the U and V tiling spinners, the dots on the column will be round and not oval.

13. Turn off Sub-Object

14. Save the scene

A Bump map does not have to be grayscale. It is very common to duplicate the map from the Diffuse Color channel into the Bump channel to add depth to a surface. You did exactly this in Chapter 3, when you dragged the bitmap of a stone wall from the Diffuse Color channel and dropped it onto the Bump channel to give the stone wall coarseness and depth. The bump effect works because of the value of each pixel in the bump map, not the hue, so a color bitmap can work quite well as a bump map. Since a color map doesn't usually have as large a difference in values as a grayscale map does, it usually needs to be used at a much higher usage strength than a grayscale map.

Opacity Map

An Opacity map uses lighter areas of the map to make the surface more opaque, and darker areas to make the surface more transparent. Values in between totally white and totally black make translucency. Use Opacity mapping to make the base of the reception counter appear to be made from sheet metal with a pattern of star-shaped holes punched through it.

1. Drag the Time Slider to frame 50

2. In the Material Editor, make the next unused sample window active

3. Name the material Star-Punch

4. Set the shader type drop-down list to Anisotropic, if it isn't already

5. In the Anisotropic Basic Parameters:

   Set Ambient and Diffuse color to: H = 120, S = 26, V = 180

   Specular Level = 50

   Glossiness = 20

   Anisotropy = 25

   Orientation = 0

6. With the camera view active, drag the sample sphere from the Material Editor and drop it onto the Reception Base, pausing to let the tooltip verify that the cursor is over the base.

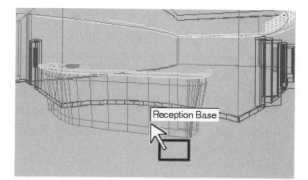

7. Quick-Render the camera view

There is an Omni light in this scene called $ReceptionBase. It is a highlight-caster, like the one for the elevator doors, set to only illuminate the Reception Base, and to only illuminate the specular areas. This light is creating the bright band on the front of the base.

8. In the Material Editor, scroll to the Maps rollout (open it if necessary)

9. Choose the button labeled None in the Opacity channel

10. In the Material / Map Browser, highlight Bitmap, choose OK

11. In the Select Bitmap Image File browser, browse to C:\Viztutorials\Chapter7, highlight StarPunch.tga, and choose Open

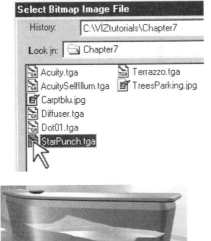

12. Render Region the area around the reception counter

Both the tiling and the mapping coordinates need to be addressed. Start with the mapping coordinates.

13. Type H, select Reception Base

14. In the Modify panel, apply a UVW Map modifier

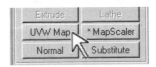

Mapping Coordinates

Adjusting mapping coordinates involves resizing and reorienting the mapping Gizmo, which can be thought of as a bitmap projector. Currently the Gizmo is a flat plane, projecting the image of the star straight down onto the counter base, at a large scale. The Gizmo needs to be stood up, to project the star sideways at the base. The base has planes that face

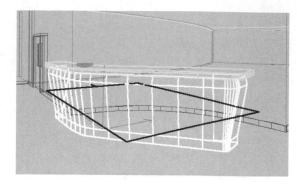

in four general directions: front, back, and sides. If the Gizmo is stood up and aimed so that the direction of projection is perpendicular to the front and back of the base (the image is like x-rays– its projection does not stop when it hits a surface), then the star image will look correct on the front and back, but will be projected nearly parallel to the sides, and won't look like stars, but likely will just show as streaks. What is needed is a box-shaped Gizmo, which will project the star image from six directions. Each plane of the counter base will pick up the projection of the star from whichever plane of the Gizmo provides the most perpendicular projection vectors.

1. In the Modify panel, set the mapping type to Box

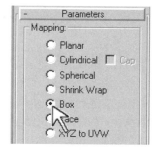

Mapping types Planar, Cylindrical, Spherical, and Box are self-explanatory. Shrink Wrap solves a problem that arises occasionally with Spherical. Imagine the spherical mapping of an image of leather onto a basketball. The image is placed at the side of the ball and wrapped completely around. Where the edges of the image meet at the opposite side of the ball, there will be a seam. In most cases the ball can be rotated so that the seam is not visible from the camera's point of view, but not always. Shrink Wrap mapping places the bitmap on top of the sphere, wraps downward, and gathers the image in a pucker at the bottom of the sphere (think of a balloon). So long as the bottom of the sphere is not visible, Shrink Wrap provides seamless mapping. However, all that wrapping, stretching, and gathering of the bitmap distorts the image. The larger the pixel dimensions of the bitmap being wrapped, the less distortion. If you place a large sphere around your building model and use Shrink Wrap to map an image of a partly cloudy sky onto the inside of the sphere to serve as a seamless sky for your panoramic Smoothmove rendering, this bitmap of sky should be large– a couple thousand pixels in its longest direction to keep the distortion to a minimum.

Face mapping maps the bitmap once onto every face of the model. It's unlikely you will ever use it. XYZ to UVW mapping is used with procedural maps; it makes them behave more like bitmaps.

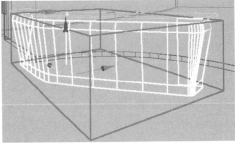

In the views you now see an orange box around the counter base. The box needs to be oriented to the base properly, and sized down.

2. With the camera view active, choose the Rotate tool, then right-click over it to show the Transform Type-In

You want to rotate the Gizmo 35 degrees. If you rotate now, you will rotate the whole counter base, not the Gizmo. The Gizmo is a sub-object of the modifier, so to rotate just the Gizmo, you need to be in sub-object mode.

3. In the Modify panel, turn on Sub-Object

4. In the Transform Type-In, enter 35 in the Offset:World Z: spinner, and hit Enter on the keyboard. Close the Transform Type-In

Now the size of the Gizmo needs to be corrected.

5. In the Modify panel, set the Length, Width, and Height of the mapping Gizmo to 1 foot

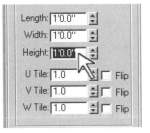

6. Render Region just the area around the reception counter

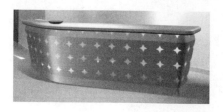

The star pattern is working now, but its scale is still too large. You could keep sizing down the mapping Gizmo to decrease the scale, but the correct size mapping Gizmo would be only a few inches, which makes it hard to see and manipulate in the views. Better to leave the Gizmo at a foot square and adjust the tiling of the bitmap in the Material Editor. Tiling, you'll recall, is the number of times a map repeats over a given distance, and in this case this distance is 1 foot.

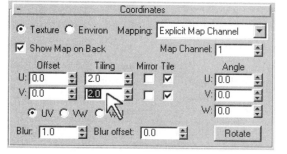

7. In the Material Editor, set both Tiling spinners to 2

The Offset spinners allow you to nudge the bitmap around on the surface.

8. Render Last

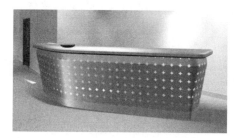

So the general approach to controlling the scale of bitmaps applied to an object is to get close with the mapping Gizmo, then fine-tune with the Tiling spinners.

The scale of the star pattern is now acceptable, but suppose you want to nudge the pattern around, maybe to perfect the way the pattern meets the closest corner. The way to do this is to make the bitmap appear on the object right in the view, so you can see the effect of changes to the Offset spinners immediately, without needing to render.

9. From the menus, choose View / Display / Shade Selected

The base becomes shaded, while the rest of the view remains in wireframe.

10. In the Material Editor, in the row of buttons just under the sample windows, choose the Show Map in Viewport button

You should now see the star pattern on the base in the camera view.

This bitmap reads reasonably well in the shaded mode, but this is not always the case. If a bitmap is distorted when shown in a shaded mode viewport, right-click over the view label, and from the pop-up menu choose Texture Correction.

11. In the Material Editor, click-hold on either of the small arrows at the right of the U: Offset spinner, and drag the mouse up and down slowly while watching the shaded base in the camera view. The star pattern will move right and left on the base.

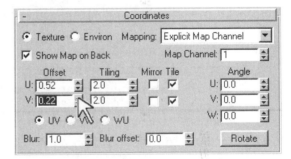

Set the U: Offset spinner to .52

Set the V: Offset spinner to .22

12. Render Last

The placement of the star pattern isn't quite perfect at the front corner. Getting it perfect is a matter of trial and error, changing settings for Tiling and Offset, and maybe opening the star bitmap in a paint program and adding a bit more white border around the black star, so the stars could be spaced a bit further apart.

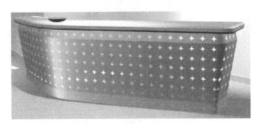

13. Save the scene

Masks

Masking is one of the most powerful tools available for precisely controlling the placement and blending of materials. With a knowledge of masks there is no material you cannot build, and without this knowledge, you will frequently run into challenges in the Material Editor that you just cannot solve. Masking involves the use of two or more maps. One of the maps is the mask, and it is generally a grayscale map. The other maps can be of any variety: Diffuse Color map, Bump map, Specular Level map, and so on.

The name Mask suggests an image you can use to understand the idea. Think of how you use masking tape when painting. Imagine that you want to repaint a wall, but you want to preserve a stenciled border that runs around the room at chair rail height. You would simply mask the border off with masking tape, and start painting. Masking in computer graphics, be it in VIZ or a paint program like Photoshop, is the same idea. The paint is the Diffuse Color map or a map in any other channel, and the masking tape is a grayscale map in which white areas are where the paint goes, black areas are masked off and will not be painted, and areas with shades of gray allow some paint to bleed through.

You'll use masking now in designing a material for a tile floor. In a single material, you need the appearance of the stone tiles, the appearance of the grout lines, and a means of controlling where stone tile appears and where grout lines appear.

The stone tile will be generated by a scanned image of brown polished stone:

The color of the grout lines will be provided by the color swatches in the material's basic parameters

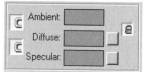

The mask that will control placement of stone tile and grout lines is a white square with a black border.

The scale at which you will apply these tiles has been greatly exaggerated for the sake of example—they are being designed as 16″ square.

1. In the Material Editor, make the next unused sample window active

2. Name the material Tile-Brown Stone

3. Set the shader type to Blinn

4. Lock the three color swatches together, and set them to:
   H = 28, S = 26, V = 150

This light brown will be the grout color.

5. Set the Specular Level to 15 and the Glossiness to 0

Setting the Glossiness to zero is a choice that is presented with a disclaimer: it is not a rule, but the author's frequent preference for walls, floors, and ceilings. Other sources will contradict this and instruct you to set glossiness to a moderate value (20 to 40) for large flat surfaces. You should try both choices and decide for yourself which looks best. The decision to lean toward a zero setting for Glossiness was the result of months of frustration in trying to get both good illumination and good contrast on walls in several projects. Bearing in mind that Glossiness is really about the diameter of the round highlight, and walls rarely have such highlights unless they are curved, a zero value for Glossiness was tried, and it seemed to make it much easier to get large flat matte surfaces to be well-lit, lively, and having good contrast at corners. If this choice works well for you, use it.

6. In the Maps channels, choose the button labeled None in the Diffuse Color channel

7. From the Material / Map Browser, select Mask

As you will see in a moment, when you choose Mask from the list, you are actually indicating that you want to control the Diffuse Color channel with two maps: a map for the diffuse color, and a mask.

395

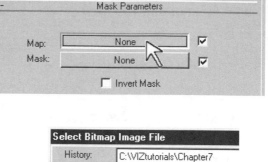

8. In the Mask Parameters,
   choose the button for the Map

9. In the Material / Map Browser,
   choose Bitmap

10. In the Select Bitmap Image File browser,
    browse to C:\Viztutorials\Chapter7 and
    choose Stone_Brn01.tga

11. Choose the Go to Parent button

12. Back at the Mask Parameters level,
    choose the button for the Mask

13. In the Material / Map Browser, choose Bitmap

14. In the Select Bitmap Image File browser, choose
    TileMask.tga

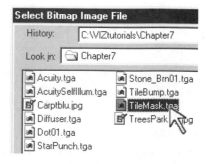

15. Choose the Go to Parent button twice, to return
    to the topmost level of the material

16. Type H, select Floor

17. In the Material Editor, choose the Assign Material to Selection button

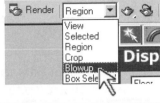

18. With the camera view active and the Time Slider at frame 50, set the render type drop-down list to Blowup and choose the leftmost Render teapot

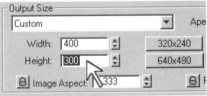

19. In the Render Design dialog box, set the Output Size to Width = 400, Height = 300, then choose Render, then choose Close to get the dialog box out of the way

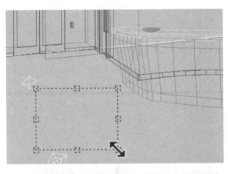

20. In the camera view, size the blowup rect-angle to render a few square feet of floor in front of the reception counter, then choose the OK button in the lower-right of the camera view

Wherever the black border of the mask bitmap prevents the application of the stone image, the diffuse area appearance is determined by the Diffuse color swatch in the Basic Parameters. So changing the color of the grout is as easy as changing the color of that color swatch.

The tiles are too flat at their edges, and the grout lines should show as being slightly recessed. The bump map that will make this happen is almost identical to the mask map, except with some blurring at the black border, so that the tile will appear to have some radius at the edges where it meets the grout lines.

Bump Map

1. In the Maps channels, set the Bump channel usage at 70, then choose the map button for the Bump channel

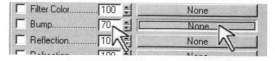

2. In the Material / Map Browser, choose Bitmap

3. In the Select Bitmap Image File dialog box, select TileBump.tga

4. Render Last

The tiles now appear to be radiused at their edges and the grout lines appear slightly sunken.

This material is beginning to become complicated. You have used maps in three different ways so far, and there are still two map usages left to complete the material. This is a good material for examining the use of the Material / Map Navigator.

Material / Map Navigator

1. At the far right of the tools below the sample windows, choose the Material / Map Navigator button

Instead of navigating up, down, and across the various levels and siblings of this material using the maps buttons, the Go to Parent button, and the Go to Sibling button, use this hierarchical tree to quickly select your desired location in the material structure.

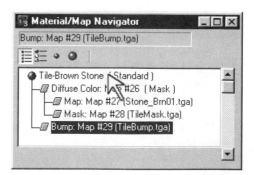

1. Choose the Tile-Brown Stone name in the Navigator to return to the top level

398

Flat Mirror Reflection Map

Reflections add tremendous realism and visual interest to a rendering, and they are not too difficult to do. There are three map types that can be used to make reflections.

Flat Mirror is for flat surfaces only. It calculates relatively quickly. It has two drawbacks. The first is that it is like putting a coat of shellac on the surface- you get great reflections but the shellac itself usually has some sheen, which brightens the material. The second is that it is prone to error. Sometimes it simply fails to render what should truly be reflected.

The Raytrace map is more accurate than Flat Mirror, and the reflections aren't added as another layer as in Flat Mirror, but are nicely blended into the colors of the material. The drawback to the Raytrace map is that it can be very slow to calculate.

Reflect / Refract is for curved surfaces. The reflections are derived by the renderer "looking out" in six directions from the object, rendering six maps, then melding the six into one map that is applied to the object using spherical mapping. While it is not as accurate as the Raytrace map and is prone to the same errors as Flat Mirror, it is much faster than Raytrace and easier to use. As the name implies, it also does well assigned to the Refraction channel, to simulate the distortions seen through a glass vase, for example.

1. In the Maps channels, set the usage spinner for the Reflection channel to 15 (a little reflection goes a long way)

2. Choose the button labeled None in the Reflection channel

3. In the Material / Map Browser, choose Flat Mirror

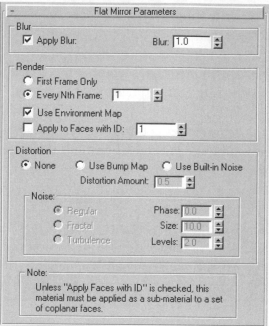

A few points about the Flat Mirror parameters (which in most cases work as they are):

If you have specified an image in the Rendering / Environment menu, you probably want that image included in the reflections, but there may be cases when you do not. In such a case, uncheck Use Environment Map.

You can distort the reflections either with the Built-In Noise (great for reflections on gently rippling water), or you can Use Bump Map, so that if you have a Noise map being used in the Bump channel of the material, this Noise pattern will not only make the object's surface appear bumpy, it will also make the reflections distort in agreement with these bumps.

At the bottom of the Flat Mirror Parameters is a note stating "unless 'Apply Faces with ID' is checked, this material must be applied as a sub-material to a set of coplanar faces." If you are applying Flat Mirror to a single flat plane, as is the case with this floor, you can ignore this note. It is directed at the attempt to apply flat mirror only to certain faces of an object, as in a cube with five wooden faces and one mirror-finish face. The face with the mirror finish needs to carry a different Material ID# than the other faces. Assume the wood faces carry Material ID 1, and the mirror face carries Material ID 2. In the Flat Mirror Parameters, Apply to Faces with ID needs to be checked, and the ID number set to 2 for the reflections to work.

4. Leave all Flat Mirror Parameters as they are, click-hold the Quick Render flyout button, and choose the green Quick Render (Production) teapot

In the Draft render settings, reflections are turned off.

5. Choose the OK button in the lower-right of the camera view

Watch the Render progress dialog box. You should see a progress bar labeled Preparing Lights, then a few moments later one labeled Rendering Reflect / Refract Maps. If you do not, cancel the rendering and select the leftmost teapot. Scroll to the Options group and make sure Auto-Reflect / Refract and Mirrors is checked, then restart the rendering.

The reflection looks great, but it has one problem. The reflections are present in the grout lines, and they should not be. The Flat Mirror map needs to be masked off where the grout lines appear. You can add masking to any map already in place.

Whether or not the grout lines reflect might seem like a trivial thing, until you compare the image before the mask, and after. Use RAM Player to compare.

6. Save the scene, then from the menus, choose Rendering / RAM Player, and choose the left teapot to load the last rendering into channel A

7. Just above the Flat Mirror Parameters rollout, at the right, choose the Type button, labeled Flat Mirror

8. In the Material / Map Browser, choose Mask

9. In the Replace Map dialog box, choose Keep old map as sub-map, and choose OK

10. In the Mask Parameters, choose the button for the Mask

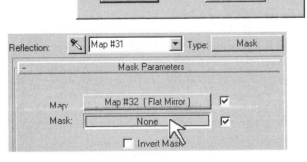

401

11. In the Material / Map Browser, choose Bitmap

12. In the Select Bitmap Image File browser, browse to the Chapter7 folder and choose TileMask.tga

13. In the Coordinates for the mask, verify that the mask is to be used as a Texture, not as an Environment

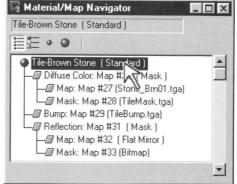

Texture versus Environment will be explained in the next section.

14. Render Last

15. In RAM Player, choose the right-hand teapot to load the last rendering into channel B, then scroll the white triangle to compare the unmasked grout lines with the masked

The masked grout lines look more realistic. This level of attention to detail is the difference between good visualizations and great ones. The grout lines are a good example of the sort of detail that your scrutiny might not consciously pick up (if it hadn't been pointed out, would you have noticed?), but which, combined with a dozen other minor discrepancies, alerts your subconscious that the image doesn't match the visual and sense memories of how such a place looks and feels. Gaining the intuition and experience to be able to discern such problems of detail and be able to fix them is the most satisfying reward for the many hours spent trying to master VIZ. Well, that and the fact that you'll be paid more.

16. In the Material / Map Navigator, choose the Tile-Brown Stone label to return to the top level of the material, then close the Navigator

It should be noted that in order to move things along in this exercise, the renaming of each level and sibling has been neglected. Actually, when using the Navigator, the default naming is fairly clear, since it is laid out in a hierarchy, but it would still help to give each level and sibling a more descriptive name.

17. Set the render type drop-down list to View, choose the leftmost teapot, choose an output size of 800 x 600, and render the entire camera view in Production mode, to see the new floor with its reflections

18. Save the scene

⚠ One more important topic on Flat Mirror reflection–troubleshooting. The sequence you just completed of first assigning Flat Mirror to the entire floor, and then going back and adding masking, was no accident. For some reason, when the Reflection channel is built as a Mask right from the start, the reflections do not render. When the Flat Mirror is added by itself, and later masked, the reflections work perfectly. This sort of behavior is not uncommon when working with Flat Mirror. If the reflections do not appear, or they appear incorrectly, your troubleshooting process should go something like this:

In the Rendering progress dialog box, did the message Rendering Reflect/Refract Maps fail to appear? If so, the problem may be in the render settings, not in the material. Select the leftmost Render teapot, and in the dialog box, make sure Auto-Reflect/Refract and Mirrors is checked.

If you see that the reflections are being calculated by the renderer, but they either appear incorrect or they don't appear at all, are you sure your surface is perfectly flat? If the surface is just a spline you drew in VIZ with a modifier applied to make it solid, then flatness is not the likely problem. But if the surface began its existence in a CAD program, or if you did any mesh editing on it, it may not be flat. Apply an Edit Mesh modifier, turn on Sub-Object Polygon, select the faces that are supposed to be flat, and in the parameters of the Edit Mesh modifier, select the Make Planar button. This technique also works well when you see long thin triangles in renderings of walls.

If flatness is not the problem, double-check the material. Try checking the Apply to Faces With ID check box, and verify in the Modify panel that the surfaces you want reflecting are assigned the proper Material ID number. If the Flat Mirror is using distortion, try turning it off. If the Flat Mirror is being masked or is part of a Mix map, try using it "straight up," as the first level with no siblings in the Reflection channel.

If none of these things makes a difference, the best strategy is to remodel the object. If possible, try a different modeling technique. Even the identical modeling technique often results in a new object that reflects properly. Flat Mirror can be quirky, and it may be faster to give up and switch to the slower and more complicated Raytrace map for the reflections on a troublesome object.

403

Reflection Map

A Reflection map is a fake reflection. For small objects at a distance, why commit rendering resources to calculating true reflections when they will not be any more convincing than if the surface just shows reflective qualities that seem to be in accordance with the environment? When rendering a high-rise building to be composited into a site photo, how do you instruct the glass to reflect the sky in the photo, when your VIZ scene contains only the building, and no environment to reflect? Reflection mapping allows you to select any bitmap image to appear as a reflection on a surface. Since no actual reflection is calculated, there is minimal effect on rendering times.

Assign a Reflection map to the railings on the second floor.

1. Drag the Time Slider to frame 54

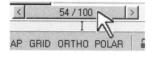

2. In the Material Editor, make the next unused sample window active

3. Name the material Steel Railing

4. Set the Basic Parameters:

    Set the shader type
    drop-down list to Metal

    Set the Ambient and Diffuse
    colors to:
    H = 142, S = 15, V = 150

    Set the Specular Level to 40

    Set the Glossiness to 70

5. Drag the material from the sample sphere and drop it onto the Railing in the camera view

6. Set the render type drop-down list to Blowup

7. Choose the leftmost teapot, and in the Render Design dialog box, set the render mode at the very lower-left to Draft, set the Output Size to the preset 640 x 480 button, verify that the Auto-Reflect / Refract and Mirrors check box is unchecked, choose the Render button, then the Close button to get the dialog box out of the way

8. Move and size the Blowup rectangle to fit around the railing, as shown here:

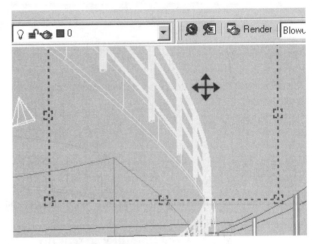

Drag the rectangle up beyond the border of the camera view, to make sure the renderer starts rendering right at the top of the view, then choose the OK button in the lower-right of the camera view

The base color of the material is dark, but when the Reflection map is added, the metal will be quite a bit brighter. There is a highlight-casting omni light in the scene, set to illuminate only the railing, and only the specular area. It is placed below and behind the railing, to suggest bounced light from a reflective upper floor.

9. In the Material Editor, set the usage of the Reflection channel to 60

10. Choose the map button In the Reflection channel

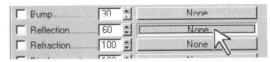

405

11. In the Material / Map Browser, choose Bitmap

12. In the Select Bitmap Image File browser, browse to the Chapter7 folder and choose House.jpg

There are two methods for mapping a bitmap or procedural map onto an object: use texture mapping, or use environment mapping. The basic difference between the two is that texture mapping is locked in position on the object, while environment mapping is locked in position relative to the environment. Imagine yourself holding a polka-dot covered soccer ball in front of a slide projector. The projector projects the word "goal" onto the soccer ball. You begin to turn the soccer ball, and the dots turn with it, but the word "goal" stays in the same place, relative to your point of view. The dots are texture mapped, the word "goal" is environment mapped. When you hold a highly reflective teapot in your hands and turn it, you will see the same items in your environment reflected on the teapot, no matter how you turn it. So reflections need to be environment mapped, and this is the default you will find in the coordinates parameters when you assign a map to the Reflection channel.

The choices for the shape of the environment that projects the reflection map onto the object are Spherical, Cylindrical, Shrink-wrap, and Screen. These choices do not necessarily have anything to do with the form of the object, they are just descriptions for the shape of the projector. The first three are conceptually identical to their counterparts in texture (UVW) mapping (see page 390). In Screen environment, it is as if the map is projected toward a rectangular screen that is perfectly perpendicular to the camera's line of sight, and fits the rectangular view exactly. Screen is the only choice that produces no distortion of the projected image. Screen generally does not work well for fake reflections, because the reflected image is too distinct and identifiable. The distortion from the other types of mapping helps disguise the fakeness of the reflection.

406

13. In the Coordinates:

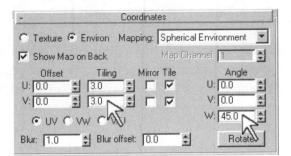

> Set both the Tiling U: and the Tiling V: to 3
>
> Set the Angle W: to 45

There is more trial and error than science to these settings. When a reflection is mapped using Spherical Environment mapping, the Tiling controls how many times the reflection map repeats across the spherical projector. Repeating only once produces very soft, indistinct reflections. Increasing the Tilings makes the reflections more distinct. Too high a Tiling will make them too distinct, and may create a discernable tiling pattern. Reflection maps often benefit from being rotated. House.jpg, which is a photo of a multihued Victorian house, has many rectilinear elements in the image, which look best when they are not aligned with the object, so the Angle W: (which corresponds to the bitmap's Z axis) is set to 45 degrees.

14. Render Last, then save the scene

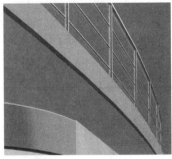

Remember that half of this effect is being accomplished by the highlight-casting omni light below and behind the railing. During development of this exercise, this light was moved several times as the material was tested.

The most crucial aspect of Reflection mapping is the map itself. It can be difficult to find an image that has acceptable colors for the scene, that shows an image that will not be readily identifiable, and that has areas of color the right size and shape to create just the reflection effect you want. If you own a digital camera, take it to a shopping mall and take pictures of anything you think will make a good reflection map. Images of partly cloudy skies make good reflection maps for small background objects like vases.

Speaking of skies and clouds, when mapping a Reflection map of a sky on the glass of a high-rise building, try using Screen mapping (assuming you are rendering a still image; screen mapped reflections do not change with point of view, so they cannot be used in animation). Because the sky is reflected most brilliantly near the top of the building, and dissipates down lower (where the hazier part of the sky near the horizon is being reflected), you should employ a gradient as a mask for the Reflection map. There is a brief exercise on this particular situation at the end of this chapter.

Raytrace Reflections

The most accurate reflections are achieved with raytracing. Raytracing can be incorporated into materials in two ways: either by designing a Standard material, and choosing Raytrace as the map type for the Reflect channel (which is what you will do in this exercise) or the Refract channel (for distortion through transparent objects), or by opening the Material Type drop-down list and choosing the Raytrace material.

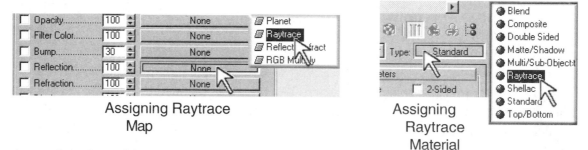

Assigning Raytrace Map

Assigning Raytrace Material

An exploration of the many and somewhat confusing settings of the Raytrace material type is beyond the scope of this book. The Raytrace map is much simpler to use than the Raytrace material, but of course it is less powerful and versatile, and the reflections it produces are not as good as those from the Raytrace material.

1. Drag the Time Slider to frame 25

2. Production Render the camera view

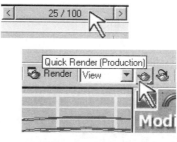

The curved glass needs to reflect the lobby and environment outside. The fact that this is an inside curve makes the reflection more challenging. If you want to see VIZ make errors in rendering, apply reflection to an inside curve. It can be made to render correctly, but the curve has to be modeled in a particular way. Examine the curved glass object in the Modify panel.

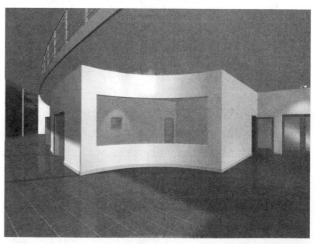

3. Type H, select Glass-Curved, then look at the
   Modify panel

The object is a NURBS surface. NURBS is another topic too
weighty to include in this book, but a quick description and a
few steps for making curved glass is manageable. The surfaces
of a NURBS object are defined by the shape and location of
curves and points, not by triangular faces arranged in 3D space. NURBS objects are
much more pliable and smooth than triangular meshes. They are perfect for curtains,
irregular plastic bottles, car bodies, and they are used extensively in character animation.

For reasons unknown, every method besides NURBS for modeling the curved glass in
this scene results in significant distortion of reflections. The specific NURBS technique
used to model this glass is called the U Loft, and it is done like this (try it if you want):

Draw an arc the shape of the glass in Top view.

Make a copy of the arc, and move it up the height of the glass.

In the Modify panel, right-click
over the Edit Stack button,
and from the menu, choose
Convert To: NURBS.

Select the other arc, repeat to convert it to a NURBS curve.

The NURBS floating toolbox should automatically open
(if not, select the NURBS Creation Toolbox button).

From the toolbox, select Create U Loft Surface.

Position the cursor over the selected arc – it will
turn blue. Pick and release. A dashed line is
attached to the cursor. Position the cursor
over the second arc, and pick. Right-click
twice to finish.

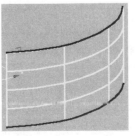

Reflections on outside curves are not so prone to error, and you can model the glass with more familiar methods, like drawing an arc and applying an Extrude modifier.

Other than the application of the Raytrace map, there's nothing unusual about a glass material with raytrace reflections. When setting up the basic parameters, the choice of shader type is important. For flat plate glass, the Blinn shader is probably the best choice. For curved glass, Blinn will work well in some situations, and Metal will be best in other scenes. Both types should be tested to find the best shader for a particular situation. The glass in this scene works best with Blinn, probably because it is an inside curve; the Metal shader makes the glass too dark.

1. In the Material Editor, make the next unused sample window active, and name the material Glass-Curved

2. Set the shader type drop-down list to Blinn

3. Lock all three color swatches together and set them to:
   H = 120, S = 20, V = 210

4. Set the Opacity to 90

You probably expect a much lower setting for opacity; 90% opaque sounds more appropriate for frosted glass than for clear. For architectural glass, the effective range of VIZ's opacity setting just seems to be weighted toward the upper end.

In the case of this inside curved glass, the Specular Level and Glossiness settings have almost no effect. There is a very minimal amount of illumination striking the glass, and it is almost impossible to make a highlight appear on an inside curve. Flat glass should also have minimal illumination, so Specular Level and Glossiness settings will have little effect on flat glass. If the glass has an outside curve, you may want to position lights to cause bright highlights, and then the settings would matter.

The reason glass should be minimally illuminated is that more light just makes the glass more milky. The impression of brightness for glass should rely almost entirely on the reflections, not on the base properties of the glass or illumination striking the glass, and there is only so much that reflection can do. Actually, glass that is light-colored, highly transparent, well-lit, and highly reflective is next to impossible in VIZ. A mastery of the Raytrace material, as opposed to the Raytrace map, will give more control over these qualities of architectural glass, but there are still limitations.

It should not surprise you that there is a light in the scene especially for the curved glass. The Omni light $CurvedGlass has a Value of only 100, and is set to only provide illumination to the diffuse area of the glass; additional specular illumination only adds milkiness. The curved glass object is excluded from any other lights in the scene that might illuminate it.

5. Open the Extended Parameters rollout

6. Drag a color swatch from the Ambient, Diffuse, and Specular colors and drop it onto the color swatch for Filter. In the Copy or Swap Colors dialog box, choose Copy

This Filter color swatch really should be up with the other color swatches in the Basic Parameters area. It is the color that travels through transparent objects, and it should be addressed any time you employ Opacity of less than 100%.

To the left of the Filter swatch is another important setting: the Amount spinner under the Falloff In/Out buttons. Falloff has to do with the angle at which you view glass. Think of a large sphere made of glass a quarter-inch thick. Looking at the center of the sphere, it is fairly transparent. Looking near the edge of the sphere, it is nearly opaque, because you are looking through the greatest thickness of glass. You can see the same thing in a drinking glass. This is what Falloff addresses. The choices for the type of Falloff are In and Out. Usually you use In, which means the inner part of the sphere is most transparent. But what if the sphere is a ball of gas? Then it looks most opaque at the center, because you are looking through the most gas, and it is most transparent at the edges, where there is the least gas, and you would need Falloff type set to Out.

You are unlikely to include glass globes and balls of gas in your visualizations, so how is Falloff used in practice? Imagine the camera looks down a long expanse of plate glass window wall. The viewing angle to the glass panels close to the camera is more perpendicular, and so the glass will be more transparent. But a hundred feet away, the viewing angle to the glass is very shallow, and the glass will probably not look transparent at all. If you design the material for this glass using just the Opacity spinner in the basic

411

parameters, and not the Amount spinner in the Falloff area, all the glass will have the same transparency, regardless of viewing angle. If you leave the Opacity spinner in the basic parameters set to 100, and instead set the transparency using the Amount spinner in the Falloff area, the transparency will decrease as the viewing angle falls off from perpendicular.

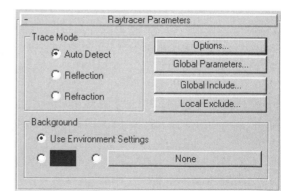

In this image, the lower glass planes use the Opacity spinner, and the far right plane is just as transparent as the far left. The upper planes use the Falloff Amt spinner. The transparency can be seen decreasing as the viewing angle falls off.

While the curved glass in the lobby scene would in reality involve a slight change in opacity between its sides and its center, the change would not be discernable, so it is acceptable to just use the Opacity spinner.

7. In the Maps channels for the Glass-Curved material, set the usage of the Reflection channel to 40. Then choose the button for the Reflection channel

8. In the Material / Map Browser, choose Raytrace

Look at the Raytracer Parameters rollout.

With Trace Mode set to Auto Detect, the renderer knows that if you are using the Raytrace map in the Reflection channel, it should calculate ray traced reflection, and if you use it in the Refraction channel, it should calculate refraction.

The Background group tells the renderer what background image or color should be incorporated into the ray traced reflections. In this case you want the trees and parking lot that are specified in the Rendering/ Environment menu to be reflected, so it should be left at Use Environment Settings. But suppose that your environment is a site photo, screen-mapped behind your building model. The ray traced reflection on the front glass facade of the building will reflect what is behind the camera– namely nothing, just blackness, because the environment is screen mapped, and only appears in front of the camera. So you would specify another bitmap for the ray traced reflections to show.

9. Choose the Options button

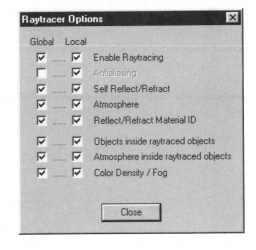

In most cases you can leave these settings at their defaults. For an explanation of each setting, see the online Help within VIZ. The Antialiasing setting bears mention. Antialiasing is the smoothing of jagged edges. With Global unchecked, no materials in the scene using a Raytrace map will employ ray tracing's special antialiasing methods. They will use the standard VIZ antialiasing methods, which usually suffice. If you want more control over how jagged edges are smoothed, or you want to be able to blur the reflections (as you might for textured glass), you will need to check Global Antialiasing.

10. Leave Global Antialiasing unchecked, and close the Options

11. Choose the Global Parameters button

If several objects in the scene use raytrace reflections, then reflections will bounce back and forth between these objects, repeatedly. The Maximum Depth spinner controls the number of bounces, and lowering it can greatly reduce rendering time. In this scene, there is only one ray traced object, so the setting is irrelevant.

12. Close the Global Raytracer Settings dialog box

13. Choose the Global Exclude button

Rarely do you need every detail of the scene expressed in reflections. You also don't want the renderer to calculate reflections for things the surface would not even reflect, because of their location or because they are obscured. This dialog box should be addressed for almost every scene, to keep rendering time manageable.

413

14. In the Exclude / Include dialog box, choose the Include button

15. From the list at left, highlight the following items:

    Ceiling, Columns, Columns02, Dish, Floor, [Glass Panels], Glass-Entry, Header-Curved, Portico Roof, Portico Slab, Reception Base, Reception Counter, Roof, and Walls

16. Choose the right-pointing arrow button to transfer the items to the window at right, then choose OK

17. Open the Antialiasing rollout and put a check in Override Global Settings

This makes these parameters available, but remember that they still will not work without the Global Antialiasing checked in the Options dialog box. Adaptive is a superior method of antialiasing. See the online Help for details. Blur blurs the entire reflection. Defocus blurs the reflection based on distance, wherein objects close to the surface reflect crisply, and objects further from the surface are blurred in the reflection.

18. Remove the check from Override Global Settings, and close the Antialiasing rollout

19. Open the Attenuation rollout

Attenuation is a similar function to Defocus, except that instead of getting fuzzier with distance, the reflection fades entirely with distance. The various choices in the Falloff type drop-down list control how quickly the reflection fades.

20. Leave Attenuation set to Off, and close the Attenuation rollout

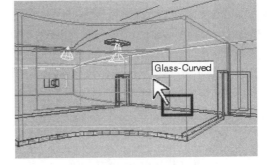

21. Drag the material from the sample sphere and in the camera view, drop it onto the curved glass, pausing to let the tooltip verify the correct object before you release

22. In either the Display panel or the Display Floater, turn on all hidden objects (there are furnishings for the conference room)

23. Render the camera view

This will take a while, maybe several minutes. You have flat mirror reflection to be calculated, you now have ray trace reflection to be calculated, and the ray trace reflection needs to be calculated twice– once during the calculation of the flat mirror, so the reflection in the curved glass will be included in the reflection on the floor, and then a second time when the renderer reaches the curved glass. Flat mirror reflections are precalculated before the scanline rendering begins, while ray trace reflections are calculated "on the fly," when the scanline renderer gets to a place in the scene where ray tracing is in use. You will see a dialog box appear at that point, labeled Raytrace Engine Setup. When the rendering is complete, notice that although the reflection in the glass was calculated for inclusion in the reflection on the floor, it is so dark you cannot even see it. Unfortunately, there does not seem to be a way to instruct the flat mirror reflection to ignore other reflections, or a way to instruct the raytrace reflection not to include itself in the flat mirror.

24. Save the scene

Decals

You know how to make a bitmap repeat over a surface, but how do you apply a bitmap just once, in a particular place, and make the background of the bitmap disappear, leaving a company logo on the door of a car, for example? It just requires the right bitmap, and a bit of finesse with mapping coordinates. Place a corporate logo on the curved header over the reception area.

1. Drag the Time Slider to frame 52

2. Render the camera view

You'll place this logo on the header, and make the black background invisible.

3. In the Material Editor, make the next unused sample window active

4. Name the material Logo-Acuity

5. Set the Basic Parameters:

> Set the shader type to Blinn
>
> Set the Ambient and Diffuse colors to H = 0, S = 0, V = 90
>
> Set Specular Level to 50
>
> Set Glossiness to 35

6. In the Maps channels, choose the map button in the Diffuse Color channel

7. In the Material / Map Browser, choose Bitmap

8. From the C:\Viztutorials\Chapter7 folder, select Acuity.tga

9. In the Coordinates rollout, verify that the bitmap is to be used as a Texture map, and that the Mapping is set to Explicit Map Channel

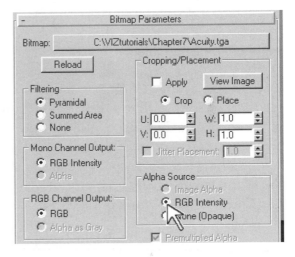

10. Uncheck both check boxes for Tile

Without Tiling, the bitmap is a decal, not a repeating texture.

11. In the Bitmap Parameters rollout, set the Alpha Source to RGB Intensity

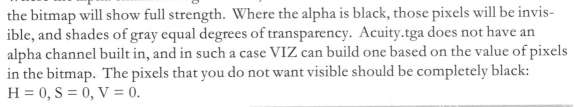

Alpha is another channel of information that can be built into a bitmap to control the transparency of the image when it is layered over other images in a paint program or used in a material in a 3D program. The alpha channel of an image is a second layer of the image, in grayscale, unseen unless you turn it on. Where the alpha channel image is white, the bitmap will show full strength. Where the alpha is black, those pixels will be invisible, and shades of gray equal degrees of transparency. Acuity.tga does not have an alpha channel built in, and in such a case VIZ can build one based on the value of pixels in the bitmap. The pixels that you do not want visible should be completely black: H = 0, S = 0, V = 0.

12. Drag the material from the Material Editor and drop it onto the curved header in the camera view

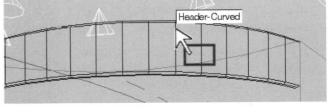

13. Render Region just the area around the header

The logo is visible, but not correctly positioned. The header needs different mapping coordinates.

Mapping Coordinates

If you were a sign painter painting this logo, you would need to get a stencil perfectly positioned on the header to begin painting. This is what you will do with the mapping coordinates – they need to be the same proportions as the bitmap, scaled to fit the height of the header, and positioned just right.

1. Type H, select Header-Curved

2. In the Modify panel apply a UVW Map modifier

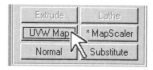

3. Turn on Sub-Object

The Sub-Object of a UVW Map modifier is the yellow and green rectangle, called the mapping Gizmo, visible in the views. If you are not in Sub-Object mode when you try to move and rotate the Gizmo, you will move or rotate the entire header.

The mapping Gizmo has oriented itself parallel to the floor, and at a rotation aligned to the World environment, not to the header object. A quick tool for aligning it to the face of the header is the Normal Align tool.

4. Near the bottom of the Modify panel, choose the Normal Align button

5. Drag the cursor on the front face of the header. The mapping Gizmo will rotate along two axes to get parallel with the face that the cursor is on. Judge the center of the header as best you can visually, and release

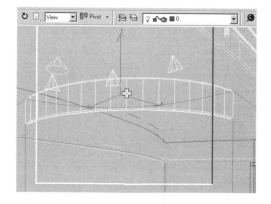

The green side of the Gizmo indicates the right-hand side of the bitmap. It is out of the view in this image, but there is a small line pointing up from the top of the Gizmo, which indicates the top of the bitmap.

6. Choose the Normal Align button again to turn it off

To get the mapping Gizmo the correct size, size its height first, then size its width to fit the proportions of the bitmap.

7. Choose the Fit button

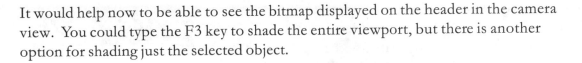

It would help now to be able to see the bitmap displayed on the header in the camera view. You could type the F3 key to shade the entire viewport, but there is another option for shading just the selected object.

8. From the menus, choose View / Display / Shade Selected

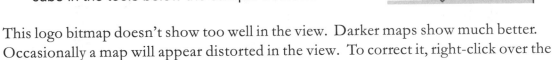

9. In the Material Editor, choose the Show Map In Viewport button. It is a blue and white checkered cube in the tools below the sample windows

This logo bitmap doesn't show too well in the view. Darker maps show much better. Occasionally a map will appear distorted in the view. To correct it, right-click over the view label and from the pop-up menu choose Texture Correction.

Showing the map on a shaded object is the best way to adjust the scale of bitmaps. For example, to get the correct scale for brick, pan and zoom one of the views to get a close-up of a door. Set the view to shaded mode, or to shade selected. Show the brick map in the view. From a reference book, look up the number of brick courses for the height of the door (standard brick runs 30 courses in 6'8"), and adjust the Length value for the UVW Map modifier (Height value if you are set to Box mapping), until you count the correct number of courses next to the door. As mentioned before, you may want to cheat the scale of brick up a bit because the correct scale often does not read well in the rendering. Once the brick is correctly scaled, you can apply a MapScaler modifier to maintain the proper scale if you change the dimensions of the wall (see the online Help for details on MapScaler).

Now set the width of the mapping Gizmo so that the Gizmo is the same proportions as the Acuity bitmap.

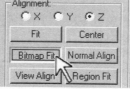

10. Choose the Bitmap Fit button

11. In the Select Image browser, select Acuity.tga from the Chapter7 folder

12. Render Last

The logo is the right scale, proportion, and is in the right position. The only problem is that it is rather dim. This is fixed with another bitmap which will control the self-illumination of the sign.

13. Turn off the Sub-Object Gizmo button, and save the scene

Self-Illumination Map

1. In the Material Editor, choose the Go to Parent button to return to the Maps channels

2. Choose the map button in the Self-Illumination channel

3. In the Material / Map Browser, choose Bitmap

4. In the Chapter7 folder, choose AcuitySelfIllum.tga

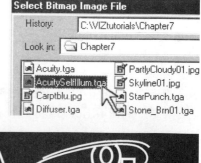

The black pixels in this bitmap will cause no self-illumination, while the white pixels will cause 100% self-illumination. The bitmap will use the same mapping coordinates as the Diffuse Color map.

5. In the Coordinates rollout, verify that the bitmap is to be used as a Texture map, and that the Mapping is set to Explicit Map Channel

6. Uncheck both check boxes for Tile

7. Render Last

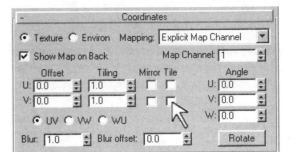

The logo reads much more clearly. The self-illumination map will also work well as a Bump map; the white pixels of the map will appear to raise the logo off the header.

Bump Map

8. Choose the Go to Parent button to return to the Maps channels.

9. Drag the Self-Illumination map button and drop it onto the Bump map button. In the Copy (Instance) Map dialog box, choose Instance

10. Set the usage of the Bump channel to 80

11. Render Last, then save the scene

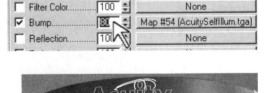

Decide which look you like best: the flush logo or the raised. If you prefer the flush, there are two ways to disable the Bump channel. You can simply uncheck the check box at the left of the Bump channel. This will leave the map assignment and all its settings, in case you want to reactivate it later. Alternately, you can drag one of the map buttons labeled None and drop it onto the Bump channel to clear it entirely.

The header is lit by two lights: $Header-L, and $Header-R. The header is excluded from all other lights. The header requires two lights because a specular highlight looks good near the right end, and the left end should be darker and have no highlight. Highlighting is kept away from the logo area, so the logo reads well against a dark background.

Double-Sided Material

The last item in the scene to design a material for is the small ornamental dish sitting on the reception counter. It will be the subject for examining the Double Sided material. The dish will get one surface treatment outside, and another inside, within the same material.

1. Type H, select the Dish

2. In either the Display panel or the Display floater, turn unselected items off, leaving only the dish displayed in the views

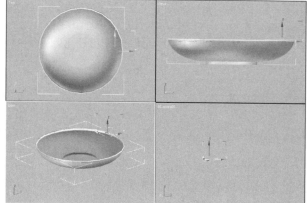

3. Zoom Extents All, then make the Left view active and Arc-Rotate to turn the Left view into a User view, showing both the outside and inside of the dish, as shown here:

4. In the Material Editor, make the next unused sample window active and name it Dish-Southwest

5. To the right of the name field, choose the Type button, currently labeled Standard

6. In the Material / Map Browser, choose Double Sided

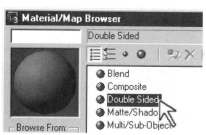

7. In the Replace Material dialog box, leave the choice at Keep old material as sub-material, and choose OK

422

8. Select the Get Material button

9. In the Material / Map Browser, set the Browse From: choice to Mtl Library

10. The browser should by default open library 3dsviz.mat. If it does not, select the File / Open button in the browser, and open 3dsviz.mat

11. Scroll to the Plastic materials, drag material Plastic-Aqua Glaze from the browser, and drop it onto the Facing Material button

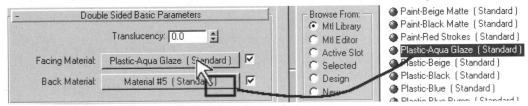

12. Scroll to the Tile materials, drag material Tile-S.W. Pattern from the browser, and drop it on the Back Material button

13. Choose the Assign Material to Selection button

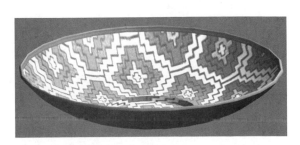

14. Render the User view

The Double-Sided material is not really a new type of material; it just joins two Standard materials in a certain way. The same goes for the Blend, the Composite, the Multi/Sub-Object, the Shellac, and the Top/Bottom material types. The only material types that don't use Standard materials are Matte/Shadow (see Chapter 2, page 83) and Raytrace.

Material Libraries

When you design good materials and want to use them in other projects, you have two choices. You can save the materials into an existing or new library, or you can use any VIZ scene as a library. In this section you will make a new library to hold the materials you've built in this chapter, then import an additional material into the library from another VIZ file.

1. If the Material / Map Browser is not still open, choose the Get Material button to open it

2. In the browser, set the Browse From: choice to Design

The list of materials now shows every material present in the scene. Notice that there are several materials listed that do not appear in the sample windows of the Material Editor. A material does not need to be in the editor to be used in the scene.

3. In the browser, choose File / Save As

4. In the Save Material Library dialog box, browse to C:\Viztutorials\Chapter7, name the library Chapter7.mat, and choose Save

As you can see in the browser, this new library contains several default materials that do not need to be included in a library. Open the library and edit them out.

5. In the Material / Map Browser, set the Browse From: choice to Mtl Library

6. In the File group, choose Open. Browse to the \Chapter7 folder, and open the Chapter7.mat library you just saved

7. Highlight the first default material in the list, and from the buttons above the list, select Delete From Library

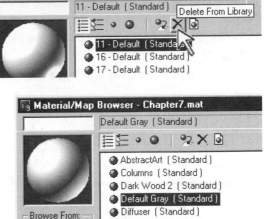

8. Repeat to delete all but one material containing the word Default. Leave one instance of Default Gray

Merge a material into this library from another scene.

9. In the Material / Map Browser, choose File / Merge

10. In the Merge Material Library dialog box, set the Files of Type drop-down list to 3D Studio VIZ (*.max)

11. Browse to C:\Viztutorials\Chapter1, highlight StillLife01.max, and select Open

12. In the Merge dialog box, highlight the Wine Bottle material, and choose OK

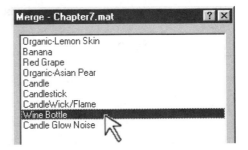

13. Choose File / Save to update the library, then close the Material / Map Browser and close the Material Editor

Smoothmove Panoramic Render

Before closing this lobby scene, generate a Smoothmove panoramic rendering of it. This will take a while, so plan to take a break while it renders.

1. In either the Display panel or the Display floater, turn on the display of all objects

2. Zoom Extents All

3. Drag the Time Slider to frame 40, and make the #Camera01 view active

4. Save the scene

5. Switch to the Utilities panel and choose Smoothmove Panoramas

6. Choose the Render button

7. Set the Output Size to the 2048 x 1024 preset button (if you have sufficient rendering time, you might set it to 3072 x 1536 for a clearer image)

8. In the Render Output group, select the Pan Files button, browse to the C:\Viztutorials\Chapter7 folder, name the file Lobby.pan, and choose Save

9. For Compression, choose JPEG Compression at high quality / large file size

10. Choose Render

The Smoothmove panorama opens showing frame 40 of the animation. Whatever the camera view shows will be the "home" view of the panorama. Drag the cursor to pan about, drag up-down with the Ctrl key held to zoom, and hit the spacebar to return to the home view. See Chapter 6, page 355, for information on downloading the free Smoothmove viewer.

426

Exterior Glass

An overview of materials would be incomplete without a quick look at a situation you will encounter in almost every exterior view: reflection of sky or streetscape on glass. A simple file has been set up to demonstrate the basic strategy for exterior reflections.

1. From the menus choose File / Open, browse to C:\Viztutorials\Chapter7, and open HighRiseGlass.max

2. Quick-Render the camera view

The stripes are not done with a bitmap, but with the Checker procedural map. The trick to turning checkers into stripes is this:

In the Coordinates, switch the U direction from Tile to Mirror. Set the U: Tiling to zero. Leave the V direction set to Tile, and set its tiling to 10. The result is ten horizontal stripes.

	Offset	Tiling	Mirror	Tile
U:	0.0	0.0	✓	
V:	0.0	10.0		✓

3. Open the Material Editor. The first material, HighRiseGlass, should be the active sample sphere

4. In the Maps channels, set the usage of the Reflection channel to 50

☐ Bump.............	30	None
☐ Reflection..........	50	None
☐ Refraction..........	100	None

5. Choose the map button for the Reflection channel, and from the Material / Map Browser, choose Mask

- Marble
- Mask
- Mix
- Noise

6. In the Mask Parameters, choose the button labeled Map

Map:	None	✓
Mask:	None	✓

7. In the Material / Map Browser, choose Bitmap

8. Browse to the Chapter7 folder and select PartlyCloudy01.jpg

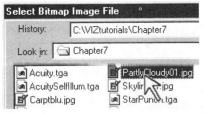

427

9. In the Coordinates for the bitmap, verify that the bitmap is to be applied using Environment mapping. Open the Mapping drop-down list and set it to Screen

Other references may suggest you use Spherical Environment mapping, and if you want the reflections to be indistinct and only suggestive of the surroundings, you should. If you want the reflections more crisp, or more identifiable as a specific scene, use Screen mapping.

10. Choose the Go to Parent button

11. Render the camera view

Blue sky and clouds reflected in a glass facade will reflect best near the top of the building, and fade toward the bottom, as the reflection starts to show the hazier and more distant part of the sky. A Gradient mask can accomplish this.

12. Choose the map button for the Mask

13. In the Material / Map Browser, choose Gradient

The gradient that will mask off the reflection will fit the height of the building and change from completely white at the top, to make the bitmap show fully, to completely black at the bottom, to make the bitmap disappear.

14. In the Coordinates for the gradient, set the map to be applied using Texture mapping

The Mapping drop-down list will automatically update to read Explicit Map Channel, meaning that the gradient will be applied according to the settings of the UVW Map modifier that has already been applied to the building.

15. Drag the Color #1 swatch and drop it over the Color #3 swatch to swap them

16. Set the Color 2 Position to .25

With the median gray color of the mask halfway up the building (.5), the clouds fade out too fast.

17. Render Last

Notice the left side of the building. The screen mapped clouds continue from the front facade to the side, which is incorrect. The easiest way to correct this is probably to assign different Material ID numbers to the two sides, and make a Multi/Sub-Object material.

Select the building and examine the UVW Map modifier applied to it. The Normal Align tool was used to stand the mapping Gizmo up against the front facade, and the Fit tool was used to fit it exactly to the facade (only the height really matters because even if the Gizmo was not as wide as the facade, the gradient would still cover the width of the facade).

18. Save the scene in the Chapter7 folder, under a different name

Summary

There are several items in the lobby scene that still do not have materials developed for them, the ceilings and the glass curtain wall being the most obvious. You have enough knowledge now to design a gypsum ceiling, with a good variance of specular level to make it realistic, or a sprayed ceiling using Speckle or Stucco procedural maps in the Bump channel. You might try a wooden ceiling for the portico roof. The glass curtain wall will need true reflections. The glass panels were modeled with this in mind, and the faster Flat Mirror reflections should work well for the glass. When settings up the Flat Mirror, turn off Use Environment Map and turn on Apply To Faces With ID #1. For additional practice, try changing the flooring from stone tiles to wooden parquet floor.

As you begin to explore the development of custom materials for your designs, it would be wise to concentrate on getting the basic parameters of the materials correct before adding mapping. A Diffuse Color map and a Bump map on a surface may distract you from seeing that the basic qualities of Specular Level and Glossiness are not correct. An image filled with well designed simple materials will be more effective than an image filled with poorly designed mapped materials.

What you did not do in this chapter is adjust lights as you developed materials, and this of course would not happen in reality. The issue of the scale and orientation of maps is not tied to lighting, but most other aspects of the Material Editor are. Before returning to the basic parameters to correct a problem with the specular qualities of an object, always stop to consider for a moment whether the problem is really in the material, or whether you are asking a material to produce an effect that the current lighting situation will simply not allow. The ability of lights to be set to include or exclude certain objects, or to be set to only illuminate the ambient or diffuse or specular areas of an object makes fine-tuning a lighting scheme much easier. There is nothing wrong with adding numerous such special-purpose lights to get your materials to do what you want.

In the first real project you do in VIZ, you will find yourself needing bitmaps not represented in any of the maps categories that ship with VIZ. The Web can be a good source of bitmaps. You might start your search at www.3d-ring.com. The 3D Ring is a series of links to many of the Web's best resources for 3D artists. There are also numerous CD-ROM collections of bitmaps. Check www.digimation.com as a starting point for finding bitmap collections for sale. The best bitmaps are ones you make yourself. If you don't have a digital camera, a good paint program, and a solid knowledge of that paint program, those are things you should work toward. You really cannot do great, original work if you cannot make your own bitmaps.

CHAPTER 8

Render Settings and Output

There are no exercises in this chapter. It is a brief reference guide to settings and strategies for output to various media, including small-format and large-format printing, videotape, CD-ROM, and slides.

Small-Format Printing

A few years ago it was necessary to send digital files to a service bureau for quality output in small-format printing. Today you can buy a $300 printer that produces quality nearly as good as the high-end desktop printers of three years ago. The paper has a great deal to do with the print quality. Glossy photo paper for inkjet printers can cost a dollar a sheet, but it is well worth it.

While it is possible to print directly from VIZ, it is not recommended. You should plan to print from a paint program, to take advantage of the color correction, image resolution, filters, and text that a paint program offers. Very seldom is an image output without some touching up in a paint program, and many visualizations owe a great deal to the paint program, likely having had all foliage, people, cars, and other entourage added "in paint." Since reflections on glass is one of the more difficult challenges in VIZ, it is common to render glass without reflection, and fake it in paint. When you work on a very large model (meaning a hundred thousand polygons or more, not large in unit dimensions), you may find that your computer simply cannot render the entire scene at one time, and you may end up rendering out portions using Render Crop, and seaming the various portions in the paint program.

In this explanation, it is assumed that the paint program being used is Photoshop 5.

Time Output: The first setting in the Render Design dialog box concerns which frames are to be rendered. Single is not the only choice you will use for still images.. When you need nine different views of a building, the best procedure is to animate the camera over nine frames, each frame showing one of the desired views. Then you can set the Time Output to Range 0 to 8, and your nine views will render overnight without needing your intervention.

Output Size: This is commonly called the pixel dimensions of the image. It is not the dots-per-inch (dpi) of the print, but the correct pixel dimensions are derived from the dpi. Assume that the physical size of the printed area on the paper you will hold in your hands is 10″ across and 7.5″ high. How many colored dots should there be for every printed inch, to get a good, clear print? The short answer is 300. 10″ x 300 dots per inch = 3,000 dots, which is the width of the pixel dimensions you need to output. For the height, multiply the width by .75. 3,000 x .75 = 2,250; 3,000 x 2,250 is a common output size.

When the image has been rendered and opened in Photoshop, it is not quite ready for printing, even if everything about the image looks great. You still need to establish the correct dpi. Notice there is no dpi setting in VIZ's Render Design dialog box. When the image is opened in Photoshop, it will open at 72 dpi- a very low resolution for printing. To set it to 300 dpi, choose, from the Photoshop menus, Image / Image Size. In the Image Size dialog box, uncheck Resample Image (this is an important step so don't overlook it). Then set the Resolution field to 300, and the Width and Height fields will automatically update to 10″ and 7.5″. Close the Image Size dialog box, save the file, and you are ready to print.

Options: If your scene uses any atmospherics such as fog or volume light, make sure Atmospherics is checked. If the scene uses Glow, that is an example of a Lens Effect, so make sure Effects is checked. If the model was built in another program, and you are just using VIZ to add lights and materials, there is a good chance the model appears to have some holes in it, from Normals being flipped the wrong way. Force 2-Sided should correct this. A better way to deal with this is in materials. A material for a problem object can be 2-Sided, leaving the rest of the scene 1-sided, and therefore rendering faster.

432

Render Output: If you don't designate a file name and location in a folder on the hard drive, there will be nothing to look at when rendering is done (and VIZ will not warn you if you've forgotten to give a name and location). When you select the Files button the Render Output File dialog box opens.

The Save as type drop-down list contains numerous formats for output. Some formats are still images and some are animated images. The most popular still-image formats are the Targa (.tga), the TIF (.tif), and the JPEG (.jpg). The .tga and .tif are uncompressed. The .jpg uses a mathematical compression scheme to make the file size on the hard drive much smaller, at a cost of some image quality (you decide quality versus compression using a sliding scale). If you will be sending the file to someone else for printing, the best bet is the .tif, as it is the most widely readable format. However, be aware that the maximum pixel dimensions of a .tif are 4000 x 4000 pixels. For larger pixel dimensions, use .tga. If the file is to be uploaded to the internet for someone else to download and print, a .jpg saved at high quality will not suffer noticeable degradation of the image, and will be a much smaller file to upload and download. The .jpg format should not be used if the image is to be edited in a paint program and resaved, since the quality of the image degrades with repeated edits and saves.

The .tga format has some unique, useful controls. You have a bits-per-pixel choice of 16, 24, or 32. If you choose 32, you are choosing to render an alpha channel embedded in the image. One of the most common uses of this is to allow a sky image to be slipped into the picture behind the geometry. If the scene environment is left black, and a 32-bit .tga is rendered, you can instruct Photoshop to use the alpha information as a mask, so that images overlayed on the rendering only fill the sky area. This allows you to quickly try many skies to get just the right look and perspective for the sky. To use this technique, turn off Pre-Multiplied Alpha. If that setting is on, the edges of the scene's geometry are blended into the background, so that when you drop in a sky in Photoshop, the geometry shows a halo around it against the sky. When rendering with a sky image assigned to the VIZ environment, turn Pre-Multiplied Alpha on. The Compress option does not degrade the image quality, because it only compresses the black background, not the color information. Checking Alpha Split will cause the alpha channel to be saved to the hard drive as a second, seperate file, as well as being embedded in the color image.

Virtual Frame Buffer: Unchecking this checkbox to turn off the display of the image as it is being rendered saves a small amount of rendering time. However, the benefit of the small savings in time seems outweighed by the reassurance of seeing the actual progress of the rendering.

VIZ Default Scanline A-Buffer: This rollout contains settings specific to the default scanline renderer. You don't have to render with the default scanline renderer. VIZ has other renderers available (found by choosing, from the menus, Tools / Options, and selecting the Rendering tab), and you can purchase other rendering engines that can plug into VIZ.

Options: These are fairly self-explanatory; you generally want Mapping, Shadows, and Auto-Reflect/Refract and Mirrors checked for your final rendering.

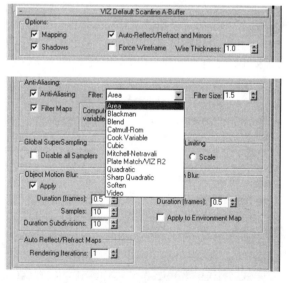

Anti-Aliasing: The smoothing of jagged diagonal lines and edges can be done with a variety of mathematical filters. For a description of each, see the online Help. VIZ's default renderer tends to soften details. If you want a more crisp image, with pronounced edges, Catmull-Rom seems to produce nice sharp images, but not excessively sharp.

Filter Maps: You can save rendering time for draft renders by turning this off, but you need it for the final render. Filter Maps is the blurring of bitmaps as they recede into the distance. Without map filtering, the grout lines in a tile floor would attempt to appear as well-defined at the far end of the hall as at the near, and the result would look bad.

Global SuperSampling: Turn it off for draft renderings, and on for final render. SuperSampling is an extra antialiasing performed on maps. It can slow rendering tremendously, but for detailed maps and maps with fine lines, it is worth the wait.

Motion Blur: These settings only pertain to animations.

Auto-Reflect/Refract maps - Iterations: It is unlikely you will need to change this, because it only pertains to the use of a Reflect / Refract map on a curved object, which is a mapping strategy seldom used in visualizations of buildings. The reflection types you will most often use are Flat Mirror and Raytrace, and this setting does not affect these types of reflections.

Watermark: Blends an image of your choice into the rendering. Use this feature to protect copyrights, to designate an image as a proof only, and to show deliverables while awaiting payment for your work.

Viewport Drop-down: You don't need to make various views active to render them. You can designate the view to be rendered right here.

Color Correction: There are no settings for this in the Render Design dialog box. It is likely that when you print, the printed colors will appear different from the colors you see when viewing the image on the monitor. VIZ has the Gamma tool that is supposed to provide a means of calibration between your monitor, your printer, and someone else's printer. If you are willing to read the online Help concerning gamma, and address the gamma settings specified by the manufacturer of your monitor and printer, it might help. If printing results in a drastic color difference, you should address gamma, as well as servicing your printer and verifying the basic settings of your monitor. Usually you can correct for color by altering the colors in Photoshop. You should make a few small versions of the image (resample the image to 320 x 240), arrange the thumbnails in another Photoshop file, make various color changes to each thumbnail, and print, to see which color changes look best when printed.

Large-Format Printing

Most of the settings that apply to small-format printing also apply to output for large-format prints. Before the final rendering, you should render a small cropped area of the scene several times, using different anti-aliasing filters, arrange the sample renders in Photoshop, and have them printed on the large printer to see which anti-aliasing will work best in that format. You should also perform different color corrections on the various samples, so the test print performs double duty.

The big issue with large-format printing is the pixel dimensions and the printed dpi of the image. You don't need high resolution for a poster-size print. Such prints are viewed from a distance, not scrutinized up close. A 150 dpi print is more than sufficient. You can get away with as low as 100 dpi, if the print is very large and the viewers will be across the room. Even rendered at pixel dimensions aimed at low dpi, images for large-format prints will have very large file sizes. To estimate how many megabytes of storage space an image file will require, multiply the pixel dimension width x the height x 3. For example, an image rendered at 640 x 480 will require just under 1 mb of storage space: 640 x 480 x 3 = 921,600 bytes. Consider, then, the file size of a poster-size print. Assuming a printed physical size of 40" x 30", and a dpi of 150, the pixel dimensions of the image are 6,000 x 4,500. 6,000 x 4,500 x 3 = 81,000,000 bytes, or 81 MB of storage space. If you attempted to render a poster-size image to be printed at 300 dpi, the file size would be 324 MB. Few computers will even open an image that large in Photoshop.

In addition to the file size issue, rendering for large-format printing may test the limits of what your computer and VIZ can handle. The larger the pixel dimensions of an image, the longer it takes to render, and the more system resources are consumed during rendering. RAM usage soars during rendering at high output size. If you will be rendering for large-format regularly, your computer needs 512 MB of RAM, and even then there will likely be some memory swapping to the hard drive on very large jobs. Attempting to render a scene containing a hundred thousand polygons at pixel dimensions of 10,000 x 7,500 can be too much for a powerful computer, especially if the scene involves lots of ray traced reflections and shadows, atmospherics like volume lights, and heavy supersampling of maps. If the computer locks up attempting the rendering job, you will have to use Render Crop (see the online Help) to render the scene in pieces, and assemble the pieces in Photoshop. If you have more than one computer available, different computers can render different areas of the image. Networked computers cannot combine processor power to join forces in calculating a single rendering.

436

Rendering Animations

Animations are rendered at lower resolutions. The rendering of each frame usually goes quickly. However, one minute of animation to be played back on videotape requires the rendering of 1,800 images which are then compiled into a single animated image. If each frame takes one minute to render, then one minute of animation will require 30 hours of rendering time. A network set up for network rendering is a huge help in this situation. One computer is the master, and in addition to rendering frames, it also monitors the other computers and assigns the next frame to whichever computer has completed its last frame and is ready for another. The more computers on the network, the shorter the rendering time.

There are two basic approaches to rendering animation: either render the entire animation into a single animated image, in one pass, or render each frame of animation as a separate file, collecting hundreds of individual files on the hard drive, then assemble all the files into a single animated image in a second rendering process. The first method is usable for short, simple animations. For animations of several hundred frames or more, the second approach is strongly recommended. If the animation is 1,000 frames in length, and you attempt to render it in one pass, only saving one animated image, and something goes wrong three-quarters of the way through the rendering (a power surge, an unexplained crash), you will have to start the process again from the beginning. If you render individual frames, you can pick up where the process was interrupted; the frames that were successfully rendered are safely stored on the hard drive.

Playback Speed and Animation Length: When setting up the animation in the VIZ scene, consider the media that will be used for playback. Different media can play animations at different frame rates (frames displayed per second). Videotape displays thirty images every second, resulting in a very smooth animated image. Some computers can accomplish thirty frames per second when the animation is playing from the hard drive. Most CD-ROM drives cannot display 30 fps, so you should assume a playback rate of 15 fps when playing from a CD. If you are creating an animation that viewers will download from the Web before viewing, you should assume 15 fps.

The animation length is derived from the playback fps. If you want one minute of animation to be played back from a CD, then 60 seconds x 15 fps = 900 frames for one minute of animation. If playback is from videotape, one minute of animation = 1,800 frames. A good strategy is to set up the animation based on playback speed of 30 fps, so it can be rendered for output to videotape. To render that animation for playback from hard drive or CD, there is a setting in the Time Output field labeled Every Nth Frame, which you would set to 2. The renderer will only render frames 0, 2, 4, 6. . ., resulting in an animation set up for 15 fps playback.

Output Size- Animation: The pixel dimensions of the window in which the animation plays is also tied to the hardware used for playback. The larger the animated image, the more pixels there are that need to be updated 15 or 30 times per second. Too many pixels overwhelm the video capabilities of a computer, and the animation either slows down or begins to skip frames to maintain the correct fps. Pixel dimensions for videotape adhere to standards that vary among countries: 720 x 486 in the United States, and 720 x 576 in Europe. Pixel dimensions for playback on computer is a judgment call. If you know that the computer on which the animation will be shown has a fast hard drive and processor, and a 32 MB video card, you might render the animation at 640 x 480. If you do not know the specifications of the playback computer, 400 x 300 is a safe output size; most CD-ROM drives can play an animation of that size at 15 fps reliably.

File Type: The two common formats for animation are .avi and .mov. The AVI format is used by Windows-based computers. The MOV format is native to the Macintosh, and requires the Quicktime player to play on a Windows-based computer. The VIZ installation screen has a choice for installing Quicktime 4.0.

When you choose the .AVI format, a dialog box appears in which you select a compression scheme for the animation. Image quality varies greatly among different compression methods, known as codecs (short for Code-Decode). Currently, the best by far of the codecs common to most computers is the Indeo video 5. This codec is directly supported by Windows 98 and Windows 2000. Windows 95

438

and NT users can download an installation program for the latest in the Indeo 5 series at www.ligos.com. If you render using Indeo 5, the playback computer must also have Indeo 5. The quality of this codec is so superior to the others that if your client does not have it installed, you should have them download and install it to view your product. Why have the end result of your dozens or even hundreds of hours of work not look its best, when downloading and installing a codec will take your client less than five minutes?

Motion Blur: This can help animations appear much more smooth. Some blurring is added to selected objects at every frame. Blurring is assigned object-by-object by selecting the object in the views and choosing the Properties menu. In the Motion Blur group, check the Enabled check box. This check box can be animated, so that motion blur for an object is only calculated during those frames when that object is moving. Object motion blur works by adding faint ghost images of an object on either side of the object, the way cartoonists depict a ball rolling. Image motion blur smears the edges of the object to give a smoother appearance of motion. See the online Help for Render Design for details. If no objects have motion blur assigned, these settings have no effect.

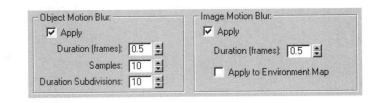

Rendering a Sequence of Stills: Page 431 mentions the strategy of rendering each frame as a still image, then compiling all the stills into an animation in a second rendering pass. Here are the steps for this strategy:

In the Render Design dialog, set the Time Output to Active Time Segment or indicate a Range of frames.

Set all options for output size, antialiasing, supersampling, motion blur, and so on.

In the Render Output group, choose the Files button. Set the Save as Type drop-down list to a still-image format; .tga or .tif are preferable. Give the files a name, and do not append a number or a suffix to the name: Ball, for example, not Ball01 or Ball.tga. Select an empty folder to save the images to. The dialog box has a Create New Folder button at the upper-right.

Render the frames of animation.

Once the rendering is finished, start a new, empty scene. Set the animation length of the new scene to the same number of frames as the scene you just rendered.

From the menus, choose Rendering / Environment

Choose the wide button labeled Environment Map. In the Material / Map Browser, choose Bitmap. In the Select Bitmap Image File browser, browse to the folder where the frames were saved. Highlight the first image in the sequence so that the name appears in the File Name field, then near the bottom of the browser check the Sequence check box. Choose Open.

The Image File List Control dialog box appears. Checking the Sequence check box tells VIZ that the background is not a single image, but a series of images that will change for every frame of animation. The Image File List (.ifl) will control the display of the images, displaying Ball0000.tga as the background for frame 0 of the animation, Ball0001.tga as the background for frame 1, and so on. Make sure that the .ifl will be placed in the same folder as the sequence of files. Leave other settings at their defaults, and choose OK.

Choose the Render Design button. Render the Active Time Segment, Set the Output Size to the same pixel dimensions as the series of stills in the .ifl. The other parameters for antialiasing, motion blur, and so on can be left at their defaults– there are no objects in the scene to which to apply these settings. Select the Files button. Set the Save as type drop-down list to the .avi or .mov format. Choose a codec. Choose Save.

Choose Render. The stills will appear in sequence as the environment background, and the end result will be a single .avi animation file.

Besides removing the requirement that the rendering process proceed from start to finish without incident to end up with a viewable file, the .ifl strategy has other benefits. Because the second rendering pass proceeds quickly, since it is just a display and compilation of previously rendered images, you have a chance to render several .avi files using different codecs at different quality settings, to find the one that gives the best compromise between image quality and file size. You might end up using two .avi files: one at highest quality (and therefore larger file size) for inclusion on a CD, and another at lesser quality, so it will download faster from the Web.

Another benefit of the .ifl strategy is that frames where you discover errors in your animation setup can often be re-rendered to fix the problem. It is very unlikely that you will set up a walkthrough of a building, render the animation, and not discover that the camera runs into a wall or that a door closes ten frames too early. If you are rendering an image sequence, you can fix the glitches in the animation and re-render just those frames, then re-render the .avi.

It is a good idea, when using the .ifl strategy, to render one .avi uncompressed, and save that .avi when you finally archive the project for storage. That way, you don't need to archive all the individual frames, which can take gigabytes. The uncompressed .avi suffers no degradation of quality, and you can retrieve it a year later to render another version of it, maybe using the latest codec, and include it on a demo reel. You'll find Full Frames (uncompressed) as one of the choices in the Compressor drop-down list.

Outputting Animation to Videotape: This requires some expensive hardware. There are numerous video boards that advertise their ability to input and output between computer and videotape, and only cost a few hundred dollars. Most of these boards are aimed at the home video market, not at professional animators. Video of children playing in the yard has much different qualities than a computer-generated walkthrough, and putting computer animation to videotape with a $300 board almost always results in disappointment. Video boards good enough for the job start at over $2,000, and you will also need to buy specialized hard drives that are not cheap. One example of such a board is the DPS Reality (www.dps.com). The DPS Perception, the precursor to the Reality, is an excellent product that is no longer in production, but you may be able to find a used one on the Web.

Service bureaus offer digital-to-tape output, and if you only plan to make videotapes a few times a year, a service might make sense financially. Make sure the service can accept your digital files directly, in their first-generation form. Many services will want to digitize and compress your file to capture it, and the quality will suffer.

Making Slides

A simple slideshow remains one of the best ways to present your images. Slide projectors are cheap, projected images look crisp and vibrant, and other than burned-out bulbs, slide projectors don't suffer from the technological glitches that can make a presentation on computer a nightmare.

You can have slides made professionally by a service bureau. The device for doing this is basically a camera loaded with slide film, mounted to a box, and aimed at a small, high-resolution monitor that displays your images. The apparatus looks very high-tech, but that's basically the concept. Talk to your service bureau about the format and resolution at which you should render images for best results in taking slides of them. Turnaround on digital-to-slides is not fast, and it is somewhat expensive.

If you need slides for tomorrow afternoon's presentation, you can get very good results taking them yourself. Get E6 type slide film (only E6 slide film can be same-day processed), and get a low ASA value, like 64. You will need a tripod and a cable release trigger. Get the best monitor in your office (the flatter the screen the better) and set it to a high resolution- 1152 x 864 or better. Your renderings should be at pixel dimensions large enough to fill most of the screen at the high resolution, although you should stay a bit away from the edges, where the screen has the most distortion. Do a few test renderings of a simple scene to get the size right. Wait until nightfall, and cover the windows as best you can to get the room as dark as possible. Don't set the tripod and camera up too close to the monitor. The closer you set up, the smaller a zoom length you will need to see the entire screen in the viewfinder, and smaller zoom length means more perspective distortion. How far away you set up will depend on the lens of your camera, but hopefully you can get about 5 feet away and zoom in. The aperture setting should be low; you need to let plenty of light in, to get good saturation on the film. You should definitely try a few different aperture settings for each image, probably bracketing in a range from your lowest setting, up to a setting of 6 or so. If you have plenty of film, try plenty of settings; you will be charged for developing the whole roll, after all, so use it up. Set the shutter speed to open for 1 second, and use the cable release to trip the shutter. If you have enough film, try a couple different shutter speeds as well.

With a good camera and monitor, a good setup, and a bit of luck, you can get quality that rivals that of the output of the fancy machines. It will cost a fraction of what the service bureaus charge, and many photo developers will have E6 slides back to you in a few hours.

442

Summary

This is graduation day. If you have worked through the seven tutorials and you have a basic grasp of the ideas and procedures presented in the exercises, then you are no longer a new user of VIZ. As an intermediate user of VIZ, you should be able to refer back to pertinent sections of this book to meet the challenges that will arise in the first few visualization projects you take on. When you no longer need to refer to the book, you can consider yourself an advanced user.

When you look at stunning images produces by seasoned users of VIZ, you probably wonder how long it will take for you to get to the level of skill that will allow you to create equally exciting renderings. There are two areas of skill that you need to develop: technical skill and artistic skill. If you practice daily and search out sources of information, you can gain a high level of technical skill in six months. Artistic skill can be more elusive. If you have never taken an art class, you really should. If you are currently enrolled in a course in computer visualization, hopefully the course includes group critiques of your images. Having your work picked apart publicly can be a bit painful, but you need to hear how your images make people feel. Your peers and your clients are going to critique your work, whether or not you are present to listen, so you may as well solicit feedback and learn from it. When you offer an honest critique of someone else's work, you will realize things about your own work that may not have occurred to you.

For the final word, one more bit of wisdom gained from experience is offered. While you need to look at what others are doing with VIZ in order to learn, don't let yourself get discouraged by how far ahead of you others may seem to be. Stunning images are meant to inspire, but they can also make you question whether you can ever get to a level of skill that will make your images equally amazing. Don't let yourself think that way. Instead, look back at images you made a year earlier and be pleased with the progress you are making. If you are asked to create a visualization that you feel is over your head, accept the challenge and don't be afraid that you lack sufficient experience. Maybe you do lack experience, but with the help of your peers you'll find a way to get the job done and you will learn in leaps and bounds.

.AVI

Audio-Video Interleaved. The most common file format for movies on a PC. VIZ renders multiple frames of the action in your scene and merges these many frames into one moving picture.

.JPG, .TGA, .TIF

Three of the most common still image file formats. These are types of bitmaps.

AEC

Stands for architecture, engineering, construction.

Antialiasing

Images are composed of many dots arranged in rows and columns. Depicting diagonal lines with such an array of dots creates a stair-step effect as the diagonal line jumps from one row or column to the next.

Antialiasing examines dots near each dot of the diagonal line and places strategically-colored dots next to the line to blur the stair-step effect. The line at left in this image uses antialiasing, the line at right does not.

Aspect Ratio

The proportions of an image. Aspect ratio is expressed either as the ratio of the width to the height (4:3), or as the value resulting from dividing the width by the height (1.333)

Bitmap

An image in digital file format. Bitmaps are composed of many dots, called pixels, arranged in rows and columns to form a picture. Synonyms for bitmap are pixel image or raster image.

Codec

Stands for Code-Decode. Codecs are compression schemes used to lower the file size of bitmaps or movies. Without a codec, the color of every pixel in an image is recorded. A codec compresses by describing the relationship between sets of pixels instead of describing every pixel. Conceptually, a codec might describe a row of pixels by saying "there is a gradient from a red pixel at the left to a green pixel at the right, over thirty pixels." In the case of a codec for a movie, the codec would describe the way one pixel changes over time: "the upper-left pixel of the movie is red at frame 1, and changes color slightly with each frame until it is green at frame thirty". Most codecs result in some loss of quality of the image.

Dpi

Stands for dots per inch. Dpi is how many dots the printer will fit into a printed inch. Dpi is synonymous with Resolution. A common dpi for an 8″ x 10″ print would be 300 dpi. For poster-size prints 150 dpi will suffice.

Map

A digital image. Colored pixels arranged in rows and columns. A map can be a bitmap, or it can be a procedural map.

Mapping Coordinates

Instructions that determine the orientation and scale of images applied onto objects in the VIZ scene. Brick is applied to a wall by applying a scanned image of brick (a bitmap) according to mapping coordinates that align the brick image properly to the wall and scale the image correctly.

Mesh

Digital fabric. A mesh is a collection of triangles arranged in 3D space to form an object. A VIZ scene is composed of one or more meshes. The word Mesh is often used as a synonym for Model.

Modifier

A set of instructions for altering an object's structure. Modifiers are packages of programming, and they remain as seperate, editable packages when they are applied to an object. Multiple modifiers can be stacked on an object to perform multiple alterations, and the parameters of any of the modifiers can be changed at any time to change the object.

Paint Program

Software for editing or creating bitmaps. Photoshop is an example of a paint program.

Pixel

A square dot. Generally the word dot is used when referring to printed images, and the word pixel is used when referring to an image on a computer monitor. If the resolution of your monitor is set to 1024 x 768, then the entirety of what you see on the screen is made up of rows of 1,024 pixels and columns of 768 pixels.

Pixel Dimensions

The number of pixels that make up a digital image. Expressed as the width times the height. Common pixel dimensions for images to be viewed on the screen are 640 x 480 or 800 x 600. Common pixel dimensions for images to be printed are much higher (in the thousands in each direction), because printers can fit many more dots into an inch than a monitor can.

Procedural Map

A digital image generated entirely mathematically. A bitmap is a file saved on the hard drive. A procedural map is not saved on the hard drive, but is generated purely mathematically at the time of rendering. VIZ offers procedural maps that can simulate the appearance of wood, smoke, water, dents, cell structure, and several other natural patterns.

Render

To render a VIZ scene is to generate a 2D colored bitmap based on the current view of the 3D scene. The 3D scene is built of thousands of triangles arranged in 3D space. Each triangle is assigned instructions for how it should look in the rendering – its hue, its brightness, its reflectivity, and so on. At rendering time these characteristics are painted on to the scene objects and a 2D bitmap image is created. That 2D image can be saved as a file on the hard drive to be enhanced in a paint program, printed, or distributed electronically.

Resolution

The number of dots per inch or the number of pixels per inch.

Transform

To determine the location, orientation, and scale of an object in 3D space. Used as both a verb and a noun. Transforms differ from modifiers in that modifiers alter the structure of an object and transforms do not – they just relocate, reorient, or resize the object.

UVW Mapping

A synonym for mapping coordinates. The letters UVW correspond to the letters XYZ used to define three directions in 3D space.

INDEX

454